THE QUIZ BOOK

MICHAEL JOSEPH
an imprint of
PENGUIN BOOKS

MICHAEL JOSEPH

UK | USA | Canada | Ireland | Australia
India | New Zealand | South Africa

Michael Joseph is part of the Penguin Random House group of companies
whose addresses can be found at global.penguinrandomhouse.com

Penguin
Random House
UK

First published 2018
001

Produced under licence from Sony Pictures Television.

Set in 9.8/11 pt Source Sans Pro
Typeset by Jouve (UK), Milton Keynes
Printed and bound in Great Britain by Clays Ltd, Elcograf S.p.A.

A CIP catalogue record for this book is available from the British Library

ISBN: 978-0-241-37888-5

www.greenpenguin.co.uk

CONTENTS

How to play

If you've ever watched *Who Wants To Be A Millionaire?* and found yourself screaming answers at the TV or thought, 'That's a piece of cake!', then brace yourselves as you take on the questions for real! With over 1,000 new questions from the writers of the show, you can put your knowledge to the test and find out who is worthy of winning the million.

Remember, answering with the correct answer is only part of the game . . . You've got to answer the quickest in the Fastest Finger First round, and then navigate through the rest of the rounds using your lifelines effectively. You can make it personal and challenge yourself, battle it out with your family on Boxing Day, or invite some friends round and find out, once and for all, who's got the brawniest brain!

For 1 Player
As on *Who Wants To Be A Millionaire?*, the aim of the game is to reach £1,000,000. Before you can even think about the cash, you must first correctly answer a question from the Fastest Finger First section. You have just 30 seconds to put the letters in the correct order. When the time's up, follow the page reference at the foot of the page to find out if you can take your place in the hot seat and begin your climb for the cash!

Once in the Hot Seat

Start with a question worth £100 and once you have decided on your final answer (and you are absolutely sure . . .) follow the page reference at the foot of the page to find out if you've won the round. If your answer is correct, you can play to win £200 and so on up the tree. The page where each money level begins is listed in the Contents section.

As on TV, you now have four lifelines, including the new 'Ask the Host' option, to help you on your way to £1,000,000. These are, of course, optional but each of them can only be used once, so use them wisely.

Fifty-Fifty

This option takes away two incorrect answers leaving the correct answer and one incorrect answer remaining. A page reference at the bottom of each page will direct you to the relevant section.

Ask the Audience

This works in exactly the same way as on *Who Wants To Be A Millionaire?*, except we've asked the audience so you don't have to! A page reference at the bottom of each page will direct you to their responses. In the end, however, the final decision is yours.

Phone a Friend

If you have a telephone handy (and a willing friend!) ring him/her up to help you out. You have thirty seconds (no cheating, now) to read the question to your friend and for them to tell you what they think the answer is. If there's someone else around, ask if they can time it for you.

Ask the Host

For the 2018 revival of the TV show, this new lifeline has been added. This gives the player in the hot seat a chance to ask the question master their thoughts on the correct answer. The host/question master has exactly one minute to share everything they know about the question and answers, including what they think to be the correct answer. After the one minute, the player decides on their final answer and the question master can turn to the answer page. As with other lifelines, the player should use this wisely as the host may not always be the best person to help!

Safety Nets

If you answer incorrectly, you are out of the game. However, there are two safety nets to help prevent anyone finishing empty handed: £1,000 is the first safety net, so if you answer a question incorrectly and you have not yet reached £1,000, not only are you out of the game, but you won't have won a penny! After you have passed the £1,000 round, and it is safely in the bank, you can set the second safety net wherever you please after you have successfully answered that question. If you have reached one (or both) of these havens and you answer a question incorrectly, then you are out of the game, but you will have won the value of the previous haven you have reached.

If at any point during the game you are unsure of an answer and don't want to risk being out of the game by answering incorrectly, you can 'stick' to the amount you have won so far and that will be your final winnings.

For 2+ Players

Players should take it in turns at being question master and posing questions to the other contestant/s. The rules are the same as for a single player. If someone reaches £1,000,000, that person is the winner and the game is over. Otherwise, whoever ends up with the most money is the winner.

Fastest Finger First

Once you have chosen your host you will need to decide who gets to go in the hot seat first, using the Fastest Finger First questions. In the absence of computers that can record players' answers to the millisecond, you can still find out who was the fastest. It's up to you whether you'd prefer for players to answer out loud or on paper. We recommend that all players write their name at the top of a post-it. When the question master has read out the question, the players have 30 seconds to write the answers in the correct order (1–4 from top to bottom) and pass to the question master. Whichever correct post-it reaches the question master first, wins.

Are you ready to play?

Good. A final piece of advice . . . With all that money at stake, think very carefully before you give your final answer. Good luck and be sure to remember the motto of *Who Wants To Be A Millionaire?* – it's only easy if you know the answer!

FASTEST FINGER FIRST

1

Going from south to north, put these clubs in order.

A: Malmo   **B:** AC Milan

C: Panathinaikos   **D:** Paris Saint-Germain

2

Starting with the nearest, put these places
in order according to their distance from London.

A: Bermuda   **B:** Belfast

C: Bucharest   **D:** Buenos Aires

3

Starting with the earliest in the year,
put these saints' days in order.

A: David   **B:** Nicholas

C: Swithin   **D:** Valentine

4

Put these Roman numerals in decreasing order of size.

A: XIX   **B:** XC

C: CM   **D:** IX

5

Starting with the youngest, put these golfers
in the order in which they were born.

A: Jack Nicklaus   **B:** Nick Faldo

C: Gary Player   **D:** Tiger Woods

Turn to the answer section on page 298 to find out if you're right

FASTEST FINGER FIRST

6

Put the four provinces of Ireland
in reverse alphabetical order.

A: Leinster

B: Connacht

C: Ulster

D: Munster

7

Starting at the North Pole and working south, put these people
in order, according to the location of their traditional homes.

A: Sherpas

B: Maoris

C: Eskimos

D: Magyars

8

Put these types of pasta in alphabetical order.

A: Rigatoni

B: Penne

C: Linguine

D: Farfalle

9

Put these countries in reverse alphabetical order.

A: Zaire

B: Zimbabwe

C: Yemen

D: Zambia

10

Starting with the fewest, put these
fairy-tale characters in numerical order.

A: Goldilocks's bears

B: Cinderella's ugly sisters

C: Dick Whittington's cat

D: Snow White's dwarfs

Turn to the answer section on page 298 to find out if you're right

FASTEST FINGER FIRST

11

Put these words in order to give the full name of the Mounties.

A: Canadian **B:** Royal

C: Police **D:** Mounted

12

Put these words in order to give the name of a 20th-century British prime minister.

A: Leonard **B:** Churchill

C: Spencer **D:** Winston

13

Put these films starring John Cleese in the order they were first released.

A: A Fish Called Wanda **B:** Monty Python's Life of Brian

C: Fierce Creatures **D:** Clockwise

14

Put these Andrew Lloyd Webber musicals in the order they were first produced on stage.

A: Jesus Christ Superstar **B:** The Beautiful Game

C: Evita **D:** Sunset Boulevard

15

Put these ghosts in the order they visit Scrooge in Charles Dickens' 'A Christmas Carol.'

A: Christmas Past **B:** Christmas Present

C: Christmas Yet To Come **D:** Jacob Marley

Turn to the answer section on page 298 to find out if you're right

FASTEST FINGER FIRST

16

Put these stages of making
a chocolate cake in the usual order.

A: Sieve the flour

B: Put into the oven

C: Decorate with chocolate

D: Allow to cool

17

Put these historical events in chronological order.

A: Battle of Britain

B: Death of Queen Victoria

C: John Paul II elected Pope

D: Suez Crisis

18

Put these European cities in order from west to east.

A: Moscow

B: Marseilles

C: Milan

D: Minsk

19

Put these famous American novels
in the order they were first published.

A: Catch 22

B: Little Women

C: The Great Gatsby

D: The Pelican Brief

20

Put these birds in alphabetical order.

A: Wood pigeon

B: Woodpecker

C: Woodchat

D: Woodcock

Turn to the answer section on page 298 to find out if you're right

FASTEST FINGER FIRST

21

Put these words in order to give
the title and name of a famous opera singer.

A: Te

B: Dame

C: Kiri

D: Kanawa

22

From north to south, put these animals in order
of the geographical area they come from.

A: Arctic fox

B: Bison

C: Emu

D: Orang-utan

23

Put these British universities in order from south to north.

A: Aston

B: Bournemouth

C: Brunel

D: Heriot-Watt

24

Starting with the most, put these plane figures in order
according to their number of sides.

A: Hexagon

B: Heptagon

C: Octagon

D: Pentagon

25

Put these boxing weights in order from heaviest to lightest.

A: Featherweight

B: Heavyweight

C: Light heavyweight

D: Middleweight

Turn to the answer section on page 298 to find out if you're right

FASTEST FINGER FIRST

26

Starting with the earliest, put these historical events in chronological order.

A: Russian Revolution

B: French Revolution

C: American Civil War

D: English Civil War

27

Starting with the lowest, put these scrabble letters in order according to their value.

A: J

B: K

C: L

D: M

28

Put these North American tourist attractions in order from east to west.

A: Grand Canyon

B: Statue of Liberty

C: Golden Gate Bridge

D: Mount Rushmore

29

Put these art movements in the order they began.

A: Impressionism

B: Art Deco

C: Renaissance

D: Art Nouveau

30

Put these blood groups in order from the most common to the rarest.

A: A

B: AB

C: B

D: O

Turn to the answer section on page 298 to find out if you're right

FASTEST FINGER FIRST

31

Put these Labour party leaders
in the order they were born.

A: Clement Attlee   **B:** James Callaghan

C: Ramsay MacDonald   **D:** Tony Blair

32

Put these people in the order they
married English or British royalty.

A: Albert of Saxe-Coburg   **B:** Catherine of Aragon

C: Philip II of Spain   **D:** Wallis Simpson

33

Starting with the fewest, put these national flags
in order of how many different colours they have.

A: France   **B:** Japan

C: United Arab Emirates   **D:** South Africa

34

Starting at the head, put these body parts in order.

A: Heart   **B:** Larynx

C: Cerebellum   **D:** Tibia

35

Put these periods of time in order from longest to shortest.

A: Fortnight   **B:** Month

C: Second   **D:** Week

Turn to the answer section on page 298 to find out if you're right

FASTEST FINGER FIRST

36

Starting with the smallest, put these pre-decimalisation coins in order of value.

A: Shilling

B: Farthing

C: Guinea

D: Crown

37

Put these in the order they are performed in a play.

A: Act 2: Scene 1

B: Act 1: Scene 2

C: Epilogue

D: Prologue

38

Put these words in order to give a traditional phrase associated with November 5th.

A: For

B: Guy

C: Penny

D: The

39

Put these years in chronological order.

A: 1000 AD

B: 100 BC

C: 10 BC

D: 1 AD

40

Put these adjectives in the order they first occur in the British National anthem.

A: Victorious

B: Noble

C: Happy

D: Gracious

Turn to the answer section on page 298 to find out if you're right

FASTEST FINGER FIRST

41

Starting with the longest, put these African rivers in order according to their length.

A: Niger | **B:** Nile

C: Zambezi | **D:** Congo

42

Starting with the shortest, put these structures in order of height.

A: Sears Tower | **B:** Eiffel Tower

C: Big Ben | **D:** CN Tower

43

Put these pioneers of aviation in the order they were born.

A: Amy Johnson | **B:** Joseph Montgolfier

C: Chuck Yeager | **D:** Orville Wright

44

Put these Asian cities in order from south to north.

A: Bangkok | **B:** Kuala Lumpur

C: Hong Kong | **D:** Tokyo

45

Put these wedding anniversaries in the order they would be celebrated by a married couple.

A: Diamond | **B:** Gold

C: Silver | **D:** Steel

Turn to the answer section on page 298 to find out if you're right

19

FASTEST FINGER FIRST

46

Put these 'Neighbours' stars in the order they had their first UK top ten hit single.

A: Jason Donovan

B: Craig McLachlan

C: Kylie Minogue

D: Holly Valance

47

Put these areas of London in alphabetical order.

A: Clapham

B: Battersea

C: Mayfair

D: Westminster

48

Starting with the most, place these countries in order by the number of others they border.

A: Spain

B: France

C: Malta

D: Portugal

49

Put the sites of these schools and colleges in order from north to south.

A: Gordonstoun

B: Harrow

C: Rugby

D: Winchester

50

Starting with the nearest, put these Islands in order according to their proximity to mainland Australia.

A: Sri Lanka

B: Borneo

C: Madagascar

D: Tasmania

Turn to the answer section on page 298 to find out if you're right

FASTEST FINGER FIRST

51

Put these Mediterranean Islands in order from west to east.

A: Corfu **B:** Corsica

C: Crete **D:** Cyprus

52

Starting with the highest, put these mountains in order.

A: Kilimanjaro **B:** Everest

C: Mont Blanc **D:** K2

53

Put these pieces of music in the order they were written.

A: Greensleeves **B:** Hey Jude

C: Lili Marlene **D:** New World Symphony

54

Put these royal houses in the order
they began to rule England or Britain.

A: Saxe-Coburg **B:** Tudor

C: Windsor **D:** York

55

Put these betting odds in order from longest to shortest.

A: Evens **B:** 4-5

C: 11-2 **D:** 33-1

Turn to the answer section on page 298 to find out if you're right

FASTEST FINGER FIRST

56

Put these words in the order they feature in the name of a charity.

A: Birds **B:** Protection

C: Royal **D:** Society

57

Put these South Americans in the order they were born.

A: Simon Bolivar **B:** Che Guevara

C: Juan Pablo Montoya **D:** Pelé

58

Starting with the shortest, put these athletics events in order of total distance covered.

A: 4 x 100m relay **B:** 800m

C: 200m **D:** 4 x 400m relay

59

Put these events in the life of Queen Victoria in chronological order.

A: Birth of first child **B:** Becomes Empress of India

C: Diamond Jubilee **D:** Death of Prince Albert

60

Put these events in Scottish history in chronological order.

A: Battle of Bannockburn **B:** Act of Union with England

C: Hadrian's Wall built **D:** Scottish Parliament opened

Turn to the answer section on page 298 to find out if you're right

FASTEST FINGER FIRST

61

Starting with the fewest, put these women in order of the number of Wimbledon singles titles they won.

A: Chris Evert

B: Martina Navratilova

C: Steffi Graf

D: Virginia Wade

62

Put these countries in alphabetical order.

A: Libya

B: Lesotho

C: Liechtenstein

D: Liberia

63

Put these items of formal wear in order from head to foot.

A: Bow tie

B: Cummerbund

C: Spats

D: Top hat

64

Put these words in alphabetical order.

A: Geology

B: Guardian

C: Gondola

D: Gallant

65

Put these men in the order they first became British prime minister.

A: Neville Chamberlain

B: John Major

C: Robert Peel

D: Robert Walpole

Turn to the answer section on page 298 to find out if you're right

FASTEST FINGER FIRST

66

Put these words in the order they appear in the title of a children's film.

A: Chocolate

B: Wonka

C: Factory

D: Willy

67

Put these people in the order they were awarded the Nobel Peace Prize.

A: Henry Kissinger

B: Nelson Mandela

C: Theodore Roosevelt

D: Albert Schweitzer

68

Put these words in alphabetical order.

A: Overstretch

B: Overpay

C: Oversensitive

D: Overtime

69

Put these words in order to give the title of a Marilyn Monroe film.

A: Hot

B: It

C: Like

D: Some

70

Starting nearest the thumb, put these fingers in order.

A: Ring finger

B: Middle finger

C: Index finger

D: Little finger

Turn to the answer section on page 298 to find out if you're right

FASTEST FINGER FIRST

71

Starting at the top of the body, put these items of clothing in the order they would normally be worn.

A: Cravat   **B:** Sneaker

C: Loincloth   **D:** Trilby

72

Put these words in order to give the name of a financial institution.

A: Exchange   **B:** New

C: Stock   **D:** York

73

Put these words in order to give the name of an international organisation.

A: Atlantic   **B:** North

C: Organisation   **D:** Treaty

74

Starting with the smallest, put these values in order.

A: Three Score   **B:** Grand

C: Ton   **D:** Baker's dozen

75

Starting with the lowest, put these darts scores in numerical order.

A: Double top   **B:** Bullseye

C: Treble 3   **D:** Double 6

Turn to the answer section on page 298 to find out if you're right

FASTEST FINGER FIRST

Put the surnames of Gail from
'Coronation Street' in chronological order.

A: Rodwell

B: McIntyre

C: Hillman

D: Platt

Put these shipping forecast areas around the
British Isles in order from north to south.

A: Forth

B: Irish sea

C: Plymouth

D: Viking

Put these words in order to give the name of a
popular Italian tourist attraction.

A: Leaning

B: Of

C: Pisa

D: Tower

Put these British seaside resorts in order from
north to south.

A: Blackpool

B: Eastbourne

C: Great Yarmouth

D: Ilfracombe

Put these human bones in order from longest to shortest.

A: Stirrup

B: Femur

C: Metatarsal

D: Tibia

Turn to the answer section on page 298 to find out if you're right

FASTEST FINGER FIRST

81

Starting with the largest, put these continents in order of size of population.

A: Africa

B: Asia

C: Antarctica

D: North America

82

Starting with the smallest, put these British coins in order of their diameter.

A: 1p

B: 2p

C: 5p

D: 10p

83

Starting with the fewest, put these words in order of how many times the letter 'z' appears.

A: Pizzazz

B: Swizzle

C: Zip

D: Zizz

84

Put these inventions in chronological order.

A: Dynamite

B: VHS recorder

C: Telescope

D: Printing press

85

Starting with the closest, put these countries in order of distance from the equator.

A: Canada

B: Ecuador

C: Ethiopia

D: South Africa

Turn to the answer section on page 298 to find out if you're right

FASTEST FINGER FIRST

86

Put these guitarists in the order they were born.

A: Eric Clapton

B: Kurt Cobain

C: Django Reinhardt

D: Bill Wyman

87

Starting with the cheapest, put these 'Monopoly' board properties in order.

A: Mayfair

B: Fleet Street

C: Regent Street

D: Old Kent Road

88

Put these words in order to make a proverb.

A: Think

B: Minds

C: Great

D: Alike

89

Put these words in the order they appear in the title of a T.S. Eliot poetry collection.

A: Cats

B: Old

C: Possum's

D: Practical

90

Put the Queen's children in the order they were first married.

A: Charles

B: Anne

C: Andrew

D: Edward

Turn to the answer section on page 298 to find out if you're right

FASTEST FINGER FIRST

91

Put these nautical distances in order from shortest to longest.

A: Cable's length

B: Fathom

C: Foot

D: Nautical mile

92

Put these cities in order from north to south.

A: Adelaide

B: Delhi

C: Vienna

D: Moscow

93

Starting at the top of the body and working down, put these conditions in order of the body part they affect.

A: Bronchitis

B: Tinnitus

C: Ingrowing toenail

D: Housemaid's knee

94

Put these south coast resorts in order from east to west.

A: Plymouth

B: Bournemouth

C: Weymouth

D: Falmouth

95

Put the answers to these sums in order from largest to smallest.

A: 12 divided by 3

B: 12 cubed

C: 12 divided by 4

D: 12 squared

Turn to the answer section on page 298 to find out if you're right

FASTEST FINGER FIRST

96

Starting with the least cooked, put these steak preparations in order.

A: Well done

B: Blue

C: Medium rare

D: Rare

97

Put these fictional characters in chronological order of their main historical settings.

A: Adrian Mole

B: Biggles

C: Scarlet O'Hara

D: Sharpe

98

Put the Teletubbies in order from largest to smallest.

A: Dipsy

B: Laa-Laa

C: Po

D: Tinky Winky

99

Starting with the lowest denomination, put these American coins in order.

A: Cent

B: Dollar

C: Dime

D: Nickel

100

Put these words in alphabetical order.

A: Overpay

B: Overtime

C: Overstretch

D: Oversensitive

Turn to the answer section on page 298 to find out if you're right

1 ◆ £100

1

If you are close to dying, you might be said to be 'at death's …'?

- **A:** Chimney
- **B:** Window
- **C:** Patio
- **D:** Door

2

Tennis players hit which of these over the net?

- **A:** Puck
- **B:** Shuttlecock
- **C:** Ball
- **D:** Discus

3

If a person gets angry about something, they are said to get hot where?

- **A:** Under the hat
- **B:** Under the collar
- **C:** Under the shirt
- **D:** Under the Y-fronts

4

Which is a British film company famous for its horror films?

- **A:** Mallet
- **B:** Screwdriver
- **C:** Nail
- **D:** Hammer

5

A rare occurrence happens once in a what?

- **A:** Green sun
- **B:** Purple star
- **C:** Blue moon
- **D:** Pink planet

If you would like to use your 50:50 please turn to page 257
To ask the audience please turn to page 278
Turn to the answer section on page 298 to find out if you've won £100!

6

What type of footwear are wellingtons?

- **A:** Trainers
- **B:** Clogs
- **C:** Slippers
- **D:** Boots

7

Which of these is a successful British band?

- **A:** Televisionfoot
- **B:** Stereoarm
- **C:** Radiohead
- **D:** Videofinger

8

Which of these is not a method of food preservation?

- **A:** Roasting
- **B:** Smoking
- **C:** Freezing
- **D:** Pickling

9

What is known as the surfing capital of South Africa?

- **A:** East London
- **B:** Sun City
- **C:** Saldana Bay
- **D:** Durban

10

Which of these is a river as opposed to an island?

- **A:** Borneo
- **B:** Amazon
- **C:** Sardinia
- **D:** Trinidad

If you would like to use your 50:50 please turn to page 257
To ask the audience please turn to page 278
Turn to the answer section on page 298 to find out if you've won £100!

1 ◆ £100

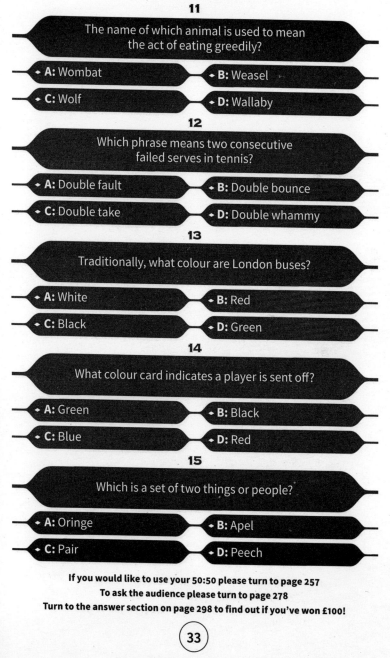

11

The name of which animal is used to mean the act of eating greedily?

A: Wombat

B: Weasel

C: Wolf

D: Wallaby

12

Which phrase means two consecutive failed serves in tennis?

A: Double fault

B: Double bounce

C: Double take

D: Double whammy

13

Traditionally, what colour are London buses?

A: White

B: Red

C: Black

D: Green

14

What colour card indicates a player is sent off?

A: Green

B: Black

C: Blue

D: Red

15

Which is a set of two things or people?

A: Oringe

B: Apel

C: Pair

D: Peech

If you would like to use your 50:50 please turn to page 257
To ask the audience please turn to page 278
Turn to the answer section on page 298 to find out if you've won £100!

1 ◆ £100

16

Complete the title of the Australian soap, 'Home and . . .'?

A: Away
B: Hurray

C: Delay
D: G'Day

17

In Britain, what is another name for a public holiday?

A: Barge holiday
B: Bank holiday

C: Barn holiday
D: Bakery holiday

18

Which of these is the name of a mountain?

A: U2
B: K2

C: R2-D2
D: CO2

19

Which of these are a type of narrow-legged trousers?

A: Peacepipe
B: Hosepipe

C: Drainpipe
D: Waterpipe

20

Which of these words is slang for money?

A: Dosh
B: Mush

C: Kudos
D: Bosh

If you would like to use your 50:50 please turn to page 257
To ask the audience please turn to page 278
Turn to the answer section on page 298 to find out if you've won £100!

1 ◆ £100

What is the name for the building
or room where soldiers eat?

A: Chaos
B: Jumble
C: Shambles
D: Mess

22

Which of these is a character from the children's
TV show 'Rainbow'?

A: Zippy
B: Buttony
C: Press-studdy
D: Safety-pinny

23

Which of these is a range of hills in England?

A: Bedswolds
B: Bunkswolds
C: Cotswolds
D: Cribswolds

24

Someone who is old-fashioned
is said to be 'behind the . . .' what?

A: Guardian
B: Times
C: Independent
D: Telegraph

25

Which of these is a spiny mammal?

A: Porcupine
B: Bakonupine
C: Hammupine
D: Sossageupine

If you would like to use your 50:50 please turn to page 257
To ask the audience please turn to page 278
Turn to the answer section on page 298 to find out if you've won £100!

1 ◆ £100

26

Which part of the Sistine Chapel did Michelangelo paint?

A: Ceiling

B: Skirting boards

C: Shelves

D: Window frames

27

Which of these phrases describes a miserly person?

A: Penny-pinching

B: Sterling-stealing

C: Lira-looting

D: Nickel-nicking

28

What are you said to get if you answer every question in an exam correctly?

A: Full matthews

B: Full marks

C: Full lukes

D: Full johns

29

Which of these is a type of beer?

A: Acid

B: Bitter

C: Tart

D: Sour

30

What was the surname of Alexander Graham, the inventor of the telephone?

A: Bell

B: Dial

C: Handset

D: Receiver

If you would like to use your 50:50 please turn to page 257
To ask the audience please turn to page 278
Turn to the answer section on page 298 to find out if you've won £100!

1 ◆ £100

What name is given to someone who is said to be
in charge but actually has little real power?

A: Figuretooth

B: Figureleg

C: Figurehead

D: Figurefoot

32

Someone who behaves oddly is said
to be 'mad as a...' what?

A: Potter

B: Hatter

C: Rotter

D: Hitter

33

On which list would a knighthood be announced?

A: Shopping list

B: Wine list

C: Laundry list

D: Honours list

34

Which of these phrases is a toast
said before having a drink?

A: Lefts up!

B: Rights up!

C: Bottoms up!

D: Tops up!

35

When something is quartered, it is divided
into how many pieces?

A: Three

B: Four

C: Twelve

D: Fifteen

If you would like to use your 50:50 please turn to page 257
To ask the audience please turn to page 278
Turn to the answer section on page 298 to find out if you've won £100!

1 ◆ £100

Which of these is an informal name for a party?

◆ **A:** Bash ◆ **B:** Crash

◆ **C:** Dash ◆ **D:** Flash

37

Which of these is an ice cream dessert?

◆ **A:** Sundae ◆ **B:** Mondae

◆ **C:** Tuesdae ◆ **D:** Wednesdae

38

A ring pull is normally used to open what?

◆ **A:** Door ◆ **B:** Can

◆ **C:** Bank account ◆ **D:** Letter

39

Something dull or ordinary is often described
by which of these phrases?

◆ **A:** Run-of-the-mill ◆ **B:** Hop-of-the-bill

◆ **C:** Skip-of-the-hill ◆ **D:** Jump-of-the-jill

40

The god Cupid is often depicted holding what?

◆ **A:** Bow and arrow ◆ **B:** Clipboard

◆ **C:** Newspaper ◆ **D:** Handbag

If you would like to use your 50:50 please turn to page 257
To ask the audience please turn to page 278
Turn to the answer section on page 299 to find out if you've won £100!

1 ◆ £100

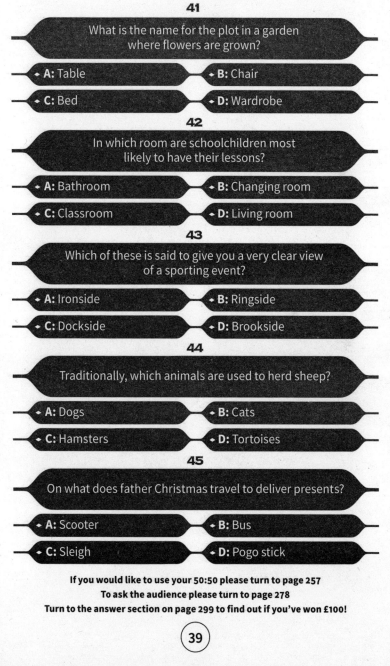

41

What is the name for the plot in a garden where flowers are grown?

A: Table

B: Chair

C: Bed

D: Wardrobe

42

In which room are schoolchildren most likely to have their lessons?

A: Bathroom

B: Changing room

C: Classroom

D: Living room

43

Which of these is said to give you a very clear view of a sporting event?

A: Ironside

B: Ringside

C: Dockside

D: Brookside

44

Traditionally, which animals are used to herd sheep?

A: Dogs

B: Cats

C: Hamsters

D: Tortoises

45

On what does father Christmas travel to deliver presents?

A: Scooter

B: Bus

C: Sleigh

D: Pogo stick

If you would like to use your 50:50 please turn to page 257
To ask the audience please turn to page 278
Turn to the answer section on page 299 to find out if you've won £100!

46

According to the saying, you should 'look before you . . .' what?

◆ **A:** Lurk

◆ **B:** Leap

◆ **C:** Lounge

◆ **D:** Lunch

47

Which of these means 'a short distance'?

◆ **A:** Stone's throw

◆ **B:** Pebble's chuck

◆ **C:** Rock's hurl

◆ **D:** Boulder's sling

48

Traditionally, what do children build on the beach?

◆ **A:** Sandflat

◆ **B:** Sandcastle

◆ **C:** Sandbungalow

◆ **D:** Sandsemi

49

Complete the title of the popular song: 'Tulips From . . .'?

◆ **A:** Rotterdam

◆ **B:** Amsterdam

◆ **C:** Aswan Dam

◆ **D:** Dontgiveadam

50

Traditionally, which of these might you find under the bed?

◆ **A:** Chamber pot

◆ **B:** Chamber of horrors

◆ **C:** Chamber orchestra

◆ **D:** Chambermaid

If you would like to use your 50:50 please turn to page 257
To ask the audience please turn to page 278
Turn to the answer section on page 299 to find out if you've won £100!

1 ◆ £100

51
What are you said to have 'in your bonnet' if you are obsessed with something?

A: Buffalo

B: Badger

C: Bee

D: Baboon

52
Which term means replacing a telephone receiver at the end of a call?

A: Jump up

B: Hang up

C: Toss up

D: Cough up

53
What is the name of the coffee specially treated to minimize preparation time?

A: Sudden coffee

B: Instant coffee

C: Abrupt coffee

D: Snappy coffee

54
If a comedian is very funny, where would the audience be said to be rolling?

A: In the aisles

B: Up the wall

C: Down your way

D: Over the top

55
Which phrase refers to charging prices that seem far too high?

A: Moonlight theft

B: Sunlight burglary

C: Daylight robbery

D: Twilight stealing

If you would like to use your 50:50 please turn to page 257
To ask the audience please turn to page 278
Turn to the answer section on page 299 to find out if you've won £100!

1 ◆ £100

56

Which of these is a programme of physical exercises?

◆ **A:** Stay able　　　　　◆ **B:** Remain trim

◆ **C:** Continue competent　◆ **D:** Keep fit

57

Traditionally, where do eloping couples marry?

◆ **A:** Bretna Blue　　　◆ **B:** Retna Red

◆ **C:** Yetna Yellow　　◆ **D:** Gretna Green

58

Which of these words describes a fairly rough sea?

◆ **A:** Cutty　　　◆ **B:** Choppy

◆ **C:** Slashy　　◆ **D:** Hacky

59

Which of these is a small animal related to the squirrel?

◆ **A:** Chipmunk　　　◆ **B:** Crispnunn

◆ **C:** Toastpreest　　◆ **D:** Nutbishup

60

Which of these is a popular television gardener?

◆ **A:** Alan Titchmarsh　　◆ **B:** Alan Titchmire

◆ **C:** Alan Titchswamp　　◆ **D:** Alan Titchbog

If you would like to use your 50:50 please turn to pages 257 and 258
To ask the audience please turn to page 278
Turn to the answer section on page 299 to find out if you've won £100!

61

In mythology, what contained all the evils of the world?

A: Pandora's box

B: Pandora's chest

C: Pandora's trunk

D: Pandora's handbag

62

Which of these is a popular garden flower?

A: Busnation

B: Carnation

C: Trainnation

D: Planenation

63

Which of these describes something
which causes stress or anxiety?

A: Knee-racking

B: Nerve-racking

C: Nodule-racking

D: Nose-racking

64

Which part of a man's suit is most likely
to be described as 'double-breasted'?

A: Tie

B: Trousers

C: Jacket

D: Waistcoat

65

What is the name of the treatment which
uses essential oils of plants?

A: Smellotherapy

B: Aromatherapy

C: Whiffatherapy

D: Pongatherapy

If you would like to use your 50:50 please turn to page 258
To ask the audience please turn to page 278
Turn to the answer section on page 299 to find out if you've won £100!

66

Which of these, added to hot water, creates a foam?

- **A:** Bubble car
- **B:** Bubble bath
- **C:** Bubble wrap
- **D:** Bubble and squeak

67

What kind of policy might cover fire and theft?

- **A:** Defiance
- **B:** Insurance
- **C:** Grievance
- **D:** Romance

68

If you want to ignore a problem, you are said to 'sweep it . . .' where?

- **A:** On top of the wardrobe
- **B:** Behind the bed
- **C:** Under the carpet
- **D:** Into the garden

69

What type of animal is 'Dumbo' in the Disney cartoon?

- **A:** Dog
- **B:** Mouse
- **C:** Elephant
- **D:** Deer

70

What is the name for putting up advertising posters in an unauthorised place?

- **A:** Ant-posting
- **B:** Bee-posting
- **C:** Fly-posting
- **D:** Wasp-posting

If you would like to use your 50:50 please turn to page 258
To ask the audience please turn to page 278
Turn to the answer section on page 299 to find out if you've won £100!

1 ◆ £100

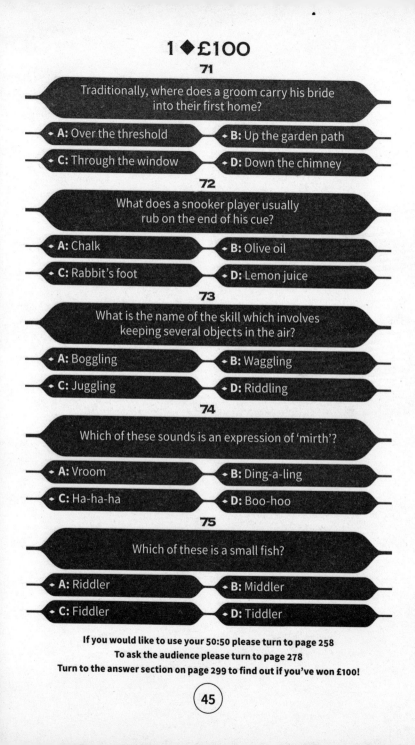

71

Traditionally, where does a groom carry his bride into their first home?

A: Over the threshold

B: Up the garden path

C: Through the window

D: Down the chimney

72

What does a snooker player usually rub on the end of his cue?

A: Chalk

B: Olive oil

C: Rabbit's foot

D: Lemon juice

73

What is the name of the skill which involves keeping several objects in the air?

A: Boggling

B: Waggling

C: Juggling

D: Riddling

74

Which of these sounds is an expression of 'mirth'?

A: Vroom

B: Ding-a-ling

C: Ha-ha-ha

D: Boo-hoo

75

Which of these is a small fish?

A: Riddler

B: Middler

C: Fiddler

D: Tiddler

If you would like to use your 50:50 please turn to page 258
To ask the audience please turn to page 278
Turn to the answer section on page 299 to find out if you've won £100!

1 ◆ £100

What is the name for the woolly ball
used to decorate a hat?

A: Babble

B: Bibble

C: Bobble

D: Bubble

77

Which of these is a sport involving horses?

A: Showleaping

B: Showhopping

C: Showrunning

D: Showjumping

78

Which word is used to describe someone
who is out of practice at doing something?

A: Fusty

B: Crusty

C: Musty

D: Rusty

79

Someone who does very well in an exam
is said to have passed with . . . what?

A: Flying colours

B: Flying fish

C: Flying doctors

D: Flying saucers

80

Something everyday and ordinary is often
described as 'common or . . .' what?

A: Greenhouse

B: Garden

C: Glasshouse

D: Growbag

If you would like to use your 50:50 please turn to page 258
To ask the audience please turn to page 278
Turn to the answer section on page 299 to find out if you've won £100!

1 ◆ £100

81

A very eccentric person is often described as being 'nutty as a . . .' what?

A: Flapjack

B: Fig roll

C: Fruitcake

D: Fondant fancy

82

Which of these would be tossed using a pitchfork?

A: Spaghetti

B: Hay

C: Caber

D: Pancake

83

What kind of envelope has a transparent section through which the address is read?

A: Fanlight envelope

B: Door envelope

C: Sunroof envelope

D: Window envelope

84

Which of these words is an expression of regret?

A: Hello

B: Thanks

C: Please

D: Sorry

85

What is the name for a prolonged period of warm weather?

A: Heat curl

B: Heat frizz

C: Heatwave

D: Heat perm

If you would like to use your 50:50 please turn to page 258
To ask the audience please turn to page 278
Turn to the answer section on page 299 to find out if you've won £100!

1 ◆ £100

On which of these might you win
a large amount of money?

A: National Flattery **B:** National Lottery

C: National Battery **D:** National Pottery

87

According to the saying, what should
you do if you can't stand the heat?

A: Live in Norway **B:** Suck a lollipop

C: Get out of the kitchen **D:** Have a cold shower

88

Which phrase refers to forcefully persuading
me to do something I'm reluctant about?

A: Twist my arm **B:** Clip my ear

C: Tweak my nose **D:** Kick my shins

89

If someone you have just been talking about suddenly
appears, you might say, 'Talk of the...' what?

A: Doctor **B:** Devil

C: Dentist **D:** Deacon

90

Which of these is the discarded end of a cigarette?

A: Dog-end **B:** Horse-end

C: Rabbit-end **D:** Stoat-end

If you would like to use your 50:50 please turn to page 258
To ask the audience please turn to page 278
Turn to the answer section on page 299 to find out if you've won £100!

91

On which of these would people
be likely to wipe their shoes?

A: Doorbell

B: Doorknob

C: Doormat

D: Doorstop

92

Which of these is a type of beer?

A: Plump

B: Podgy

C: Stout

D: Stocky

93

What name is given to a dishonest or careless workman?

A: Cowbell

B: Cowboy

C: Cowherd

D: Cowpat

94

Who would be most likely to carry a truncheon?

A: Police officer

B: Brain surgeon

C: Tax inspector

D: Landscape gardener

95

Which of these is most likely to be marketed as 'lead-free'?

A: Petrol

B: Milk

C: Hand cream

D: Coffee

If you would like to use your 50:50 please turn to page 258
To ask the audience please turn to page 278
Turn to the answer section on page 299 to find out if you've won £100!

96

Something almost impossible to find is said
to be like a needle . . . where?

A: In a department store

B: In a bird's nest

C: In a haystack

D: In a forest

97

Which of these phrases describes
a restless or fidgety person?

A: Bats in his hats

B: Goats in his coats

C: Ants in his pants

D: Dogs in his clogs

98

What is the usual name for frozen rain?

A: Hail

B: Mail

C: Pail

D: Sail

99

Which of these is a means of keeping
a farm animal in its field?

A: Electric drill

B: Electric fence

C: Electric guitar

D: Electric eel

100

Which of these phrases is one most likely
to say if caught committing a crime?

A: It's a fair cop

B: It's a nice rozzer

C: It's a lovely sergeant

D: It's a bonny bobby

If you would like to use your 50:50 please turn to page 258
To ask the audience please turn to page 278
Turn to the answer section on page 299 to find out if you've won £100!

2◆£200

1

Earth, Wind and Fire collaborated with The Emotions in 1979 to bring us which of the following?

- **A:** Boogie Wonderland
- **B:** Disco Hinterland
- **C:** Rocking Borderland
- **D:** Limboland

2

Someone who completely disappears is said to 'Vanish into . . .' what?

- **A:** Hot air
- **B:** Fresh air
- **C:** Thin air
- **D:** Bel air

3

Who is Noddy's best friend?

- **A:** Big-Nose
- **B:** Big-Feet
- **C:** Big-Ears
- **D:** Big-Mouth

4

What term is often used in the news in the context of a recession

- **A:** Sherbet dip
- **B:** Skinny dip
- **C:** Cheese dip
- **D:** Double dip

5

What is normally stored in an arsenal?

- **A:** Wheat
- **B:** Wool
- **C:** Weapons
- **D:** Whistles

If you would like to use your 50:50 please turn to page 258
To ask the audience please turn to page 278
Turn to the answer section on page 299 to find out if you've won £200!

2 ♦ £200

6

What is a male duck called?

- **A:** Drake
- **B:** Rooster
- **C:** Tom
- **D:** Cob

7

What type of vehicle is used to move goods on pallets?

- **A:** Knifelift truck
- **B:** Forklift truck
- **C:** Spoonlift truck
- **D:** Napkinlift truck

8

According to the title of a Strauss waltz, what colour is the Danube?

- **A:** Green
- **B:** Black
- **C:** Blue
- **D:** Pink

9

In the Bible, who killed Abel?

- **A:** Kevin
- **B:** Kris
- **C:** Cain
- **D:** Connor

10

Which World Cup winners are nicknamed the 'Samba Boys'?

- **A:** England
- **B:** France
- **C:** Brazil
- **D:** Germany

If you would like to use your 50:50 please turn to page 258
To ask the audience please turn to page 278
Turn to the answer section on page 299 to find out if you've won £200!

2 ◆ £200

11

Which word can mean a Scandinavian or a vegetable?

◆ **A:** Finn
◆ **B:** Dane
◆ **C:** Swede
◆ **D:** Norwegian

12

Which of these phrases means 'Be quiet'?

◆ **A:** Buckle up
◆ **B:** Braces up
◆ **C:** Bag up
◆ **D:** Belt up

13

Which of these words describes something done quickly and carelessly?

◆ **A:** Smackcrash
◆ **B:** Swatclash
◆ **C:** Spankbash
◆ **D:** Slapdash

14

What name is given to a small police patrol car?

◆ **A:** Koala car
◆ **B:** Teddy car
◆ **C:** Panda car
◆ **D:** Polar car

15

What name is given to the wife of a duke?

◆ **A:** Baroness
◆ **B:** Duchess
◆ **C:** Marchioness
◆ **D:** Lady

If you would like to use your 50:50 please turn to pages 258 and 259
To ask the audience please turn to pages 278 and 279
Turn to the answer section on page 299 to find out if you've won £200!

2♦£200

16

What best describes the Wall Street area of New York?

A: Financial

B: Fashion

C: Media

D: Entertainment

17

In motor racing what does a chequered flag signal?

A: Safety car ahead

B: Oil on track

C: Accident ahead

D: End of the race

18

Which king had two of his wives executed?

A: Macbeth

B: Richard III

C: Edward II

D: Henry VIII

19

Complete the title of the BBC sitcom: 'Men Behaving . . .'?

A: Badly

B: Awfully

C: Shockingly

D: Dreadfully

20

Confetti is most associated with which event?

A: Christening

B: Funeral

C: Hogmanay

D: Wedding

If you would like to use your 50:50 please turn to page 259
To ask the audience please turn to page 279
Turn to the answer section on page 299 to find out if you've won £200!

2 ◆ £200

21

Complete the title of the TV show: 'Top of the . . .'?

- **A:** Pops
- **B:** Mops
- **C:** Sops
- **D:** Tops

22

What are you said to take if you support someone in an argument?

- **A:** Advantage
- **B:** Leave
- **C:** Heart
- **D:** Sides

23

For which of these activities might an abacus be used?

- **A:** Counting
- **B:** Cooking
- **C:** Climbing
- **D:** Cleaning

24

Which of these is a fish-eating mammal?

- **A:** Atter
- **B:** Etter
- **C:** Otter
- **D:** Utter

25

If you 'floss', you clean between what?

- **A:** Teeth
- **B:** Toes
- **C:** Nostrils
- **D:** Ears

If you would like to use your 50:50 please turn to page 259
To ask the audience please turn to page 279
Turn to the answer section on page 299 to find out if you've won £200!

2♦£200

Which of these is a sign of the zodiac?

A: Scorpio

B: Gemino

C: Taurio

D: Caprio

27

Complete the title of the popular children's programme: 'Postman . . .'?

A: Sam

B: Ted

C: Pat

D: Bob

28

Which of these is a surprise or disappointment?

A: Bombshell

B: Eggshell

C: Seashell

D: Tortoiseshell

29

Which of these would you sniff if you felt faint?

A: Reeking vinegars

B: Smelling salts

C: Stinking mustards

D: Ponging peppers

30

Which of these is another name for embroidery?

A: Hookwork

B: Skewerwork

C: Thornwork

D: Needlework

If you would like to use your 50:50 please turn to page 259
To ask the audience please turn to page 279
Turn to the answer section on page 299 to find out if you've won £200!

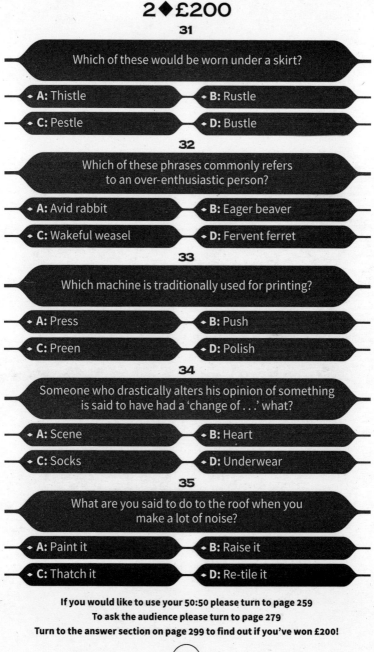

2 ◆ £200

31

Which of these would be worn under a skirt?

- **A:** Thistle
- **B:** Rustle
- **C:** Pestle
- **D:** Bustle

32

Which of these phrases commonly refers to an over-enthusiastic person?

- **A:** Avid rabbit
- **B:** Eager beaver
- **C:** Wakeful weasel
- **D:** Fervent ferret

33

Which machine is traditionally used for printing?

- **A:** Press
- **B:** Push
- **C:** Preen
- **D:** Polish

34

Someone who drastically alters his opinion of something is said to have had a 'change of . . .' what?

- **A:** Scene
- **B:** Heart
- **C:** Socks
- **D:** Underwear

35

What are you said to do to the roof when you make a lot of noise?

- **A:** Paint it
- **B:** Raise it
- **C:** Thatch it
- **D:** Re-tile it

If you would like to use your 50:50 please turn to page 259
To ask the audience please turn to page 279
Turn to the answer section on page 299 to find out if you've won £200!

36

What would you be most likely
to find on top of an Irish coffee?

A: Cream

B: Mayonnaise

C: Butter

D: Doughnut

37

The OFT is the Office of . . . what?

A: Fish Tinning

B: Fair Trading

C: Fluffy Toys

D: Fun Things

38

In which method of egg preparation are the eggs
beaten before being cooked?

A: Boiled

B: Fried

C: Poached

D: Scrambled

39

Which phrase describes something which is as good as new?

A: Chervil situation

B: Parsley position

C: Tarragon state

D: Mint condition

40

What is another name for a spy?

A: Secretaire

B: Secret agent

C: Secretary

D: Secretion

If you would like to use your 50:50 please turn to page 259
To ask the audience please turn to page 279
Turn to the answer section on page 299 to find out if you've won £200!

41

Which of these is a tonic or something which makes you feel more cheerful?

- **A:** Pick-and-mix
- **B:** Pickaxe
- **C:** Pick-me-up
- **D:** Pickpocket

42

What is the term for autumn?

- **A:** Droop
- **B:** Fall
- **C:** Fade
- **D:** Slump

43

Which of these is someone who assists at a baby's birth?

- **A:** Midwife
- **B:** Partwife
- **C:** Semiwife
- **D:** Bitwife

44

Complete the name of the great Renaissance artist: Leonardo da . . .?

- **A:** Binci
- **B:** Pinci
- **C:** Vinci
- **D:** Zinci

45

What does a person use to prevent sweat?

- **A:** Anticyclone
- **B:** Antifreeze
- **C:** Antiperspirant
- **D:** Antimatter

If you would like to use your 50:50 please turn to page 259
To ask the audience please turn to page 279
Turn to the answer section on page 299 to find out if you've won £200!

2 ♦ £200

46

Which of these phrases refers to the herding of animals before they are put into a pen?

- **A:** Squaring up
- **B:** Rounding up
- **C:** Ovalling up
- **D:** Triangling up

47

At which ceremony does a person affirm the vows made for him at his christening?

- **A:** Confirmation
- **B:** Complication
- **C:** Confrontation
- **D:** Commiseration

48

Which is a name given to an astronomical feature which has intense gravitational power?

- **A:** Black hole
- **B:** Blue pit
- **C:** White abyss
- **D:** Green crater

49

Which of these is the name of a church service?

- **A:** Flatsong
- **B:** Levelsong
- **C:** Evensong
- **D:** Smoothsong

50

With which activity is the phrase 'Going, going, gone' most associated?

- **A:** Auction
- **B:** Gardening
- **C:** Horse racing
- **D:** Boxing

If you would like to use your 50:50 please turn to page 259
To ask the audience please turn to page 279
Turn to the answer section on page 299 to find out if you've won £200!

2 ♦ £200

51

Which of these people would be most likely to wear a 'hard hat' in their work?

A: Builder

B: Doctor

C: Teacher

D: Vicar

52

A comfortable life, or situation, is often described as a bed of what?

A: Hyacinths

B: Daffodils

C: Roses

D: Dahlias

53

Someone who narrowly escapes from a difficult situation is said to be 'saved by the . . .' what?

A: Bat

B: Ball

C: Boat

D: Bell

54

Which group of people are known colloquially as 'boys in blue'?

A: Scouts

B: Estate agents

C: Traffic wardens

D: Police

55

Which of these refers to a type of puzzle?

A: Brainpower

B: Brainstorm

C: Brain-teaser

D: Brainwave

If you would like to use your 50:50 please turn to page 259
To ask the audience please turn to page 279
Turn to the answer section on page 299 to find out if you've won £200!

2 ♦ £200

56

What is the usual name for the railway carriage where drinks and light snacks can be bought?

- **A:** Breakfast car
- **B:** Buffet car
- **C:** Brunch car
- **D:** Banquet car

57

What would you be most likely to do with Cabernet Sauvignon?

- **A:** Drink it
- **B:** Clean the sink
- **C:** Groom the dog
- **D:** Polish the car

58

Which part of a ship shares its name with a collection of playing cards?

- **A:** Sail
- **B:** Anchor
- **C:** Deck
- **D:** Funnel

59

Which card game involves shouting when two identical cards are turned over?

- **A:** Bridge
- **B:** Poker
- **C:** Canasta
- **D:** Snap

60

Which of these is a spring month in Britain?

- **A:** April
- **B:** June
- **C:** September
- **D:** January

If you would like to use your 50:50 please turn to page 259
To ask the audience please turn to page 279
Turn to the answer section on page 299 to find out if you've won £200!

2 ◆ £200

61

What name is given to a pre-recorded laughter track sometimes added to TV shows?

A: Canned laughter

B: Bottled laughter

C: Cured laughter

D: Bagged laughter

62

A slice of which of these fruits is most likely to be put in a cup of tea, instead of milk?

A: Grape

B: Pineapple

C: Lemon

D: Peach

63

A person who acts in a reckless manner is said to 'throw caution to the . . .' what?

A: Sun

B: Wind

C: Snow

D: Rain

64

Which of these appears on the skin as the result of a blow?

A: Mole

B: Birthmark

C: Freckle

D: Bruise

65

What are the small trinkets often worn on bracelets?

A: Charms

B: Delights

C: Attractions

D: Appeals

If you would like to use your 50:50 please turn to page 259
To ask the audience please turn to page 279
Turn to the answer section on page 299 to find out if you've won £200!

66

Which household object is most
likely to have a hob on top?

A: Piano

B: Cooker

C: Wardrobe

D: Sofa

67

In what kind of chair do babies sit while being fed?

A: High

B: Low

C: Wide

D: Narrow

68

According to the saying, what is as good as a rest?

A: A cup of tea

B: A change

C: A box of chocolates

D: A day's shopping

69

'Sarnie' is an informal name for what kind of snack?

A: Sausage roll

B: Scotch egg

C: Sandwich

D: Soup

70

Trafalgar Square is most associated with what kind of bird?

A: Raven

B: Pigeon

C: Hawk

D: Sparrow

If you would like to use your 50:50 please turn to page 259
To ask the audience please turn to page 279
Turn to the answer section on page 299 to find out if you've won £200!

2 ◆ £200

71

Which of these phrases refers to the upper levels of government?

A: Foyers of power

B: Verandahs of power

C: Corridors of power

D: Balconies of power

72

Complete the proverb: 'Don't count your chickens before they are . . .'?

A: Plucked

B: Hatched

C: Eaten

D: Boiled

73

Which of these might be caused by a blow to the head?

A: Condensation

B: Concussion

C: Conduction

D: Confection

74

Traditionally, a croissant is eaten at which meal?

A: Breakfast

B: Lunch

C: Tea

D: Dinner

75

What is the name for the beams, often made of steel, used in building bridges?

A: Gargles

B: Girders

C: Gordons

D: Gurgles

If you would like to use your 50:50 please turn to pages 259 and 260

To ask the audience please turn to pages 279 and 280

Turn to the answer section on page 300 to find out if you've won £200!

2 ◆ £200

76

What is the nickname of a lively party, especially if dancing is involved?

A: Toes-down

B: Heads-in

C: Arms-round

D: Knees-up

77

Who would you be putting your trust in if you were 'under the knife'?

A: Surgeon

B: Carvery chef

C: Fencer

D: Sword swallower

78

Complete the first line of the classic Irving Berlin song: 'I'm Dreaming of . . .' what?

A: A lottery win

B: A summer holiday

C: A white Christmas

D: A tax rebate

79

What is a group of stables with living accomodation above, now usually converted to desirable residences?

A: Barks

B: Yelps

C: Mews

D: Squeaks

80

Which film director was known as 'The Master of Suspense'?

A: Billy Wilder

B: David Lean

C: Stanley Kubrick

D: Alfred Hitchcock

If you would like to use your 50:50 please turn to page 260
To ask the audience please turn to page 280
Turn to the answer section on page 300 to find out if you've won £200!

2 ◆ £200

81

Traditionally, who is the assistant to the bridegroom on his wedding day?

- **A:** Best man
- **B:** Smartest friend
- **C:** Neatest buddy
- **D:** Oldest brother

82

What are you said to 'dish' if you reveal scandal about someone?

- **A:** The dirt
- **B:** The grime
- **C:** The mud
- **D:** The sludge

83

What is the name for the hard deposit found in kettles?

- **A:** Limescale
- **B:** Lemonscale
- **C:** Orangescale
- **D:** Grapefruitscale

84

Which of these is a traditional accompaniment to roast turkey at Christmas?

- **A:** Wadding
- **B:** Lagging
- **C:** Stuffing
- **D:** Lining

85

What is the name for the piece of paper given by a seller to a buyer as an acknowledgement of payment?

- **A:** Counterfoil
- **B:** Receipt
- **C:** Statement
- **D:** Portfolio

If you would like to use your 50:50 please turn to page 260
To ask the audience please turn to page 280
Turn to the answer section on page 300 to find out if you've won £200!

86

Which of these words does not mean 'friend'?

A: Chum   **B:** Foe

C: Mate   **D:** Pal

87

Something which uses the most up to date technology is often described as 'cutting...' what?

A: Edge   **B:** Glass

C: Teeth   **D:** Corners

88

Which of these describes a realistic, unpretentious person?

A: Downtrodden   **B:** Downcast

C: Down-to-earth   **D:** Down-in-the-mouth

89

Which of these tools is most often used for clearing a blocked sink?

A: Clamp   **B:** File

C: Plunger   **D:** Chisel

90

What kind of item is most likely to be protected by a dust jacket?

A: Book   **B:** Milk bottle

C: Vase   **D:** Water tank

If you would like to use your 50:50 please turn to page 260
To ask the audience please turn to page 280
Turn to the answer section on page 300 to find out if you've won £200!

2 ♦ £200

91

What kind of restaurant
is most likely to serve polenta?

A: Italian   **B:** Chinese

C: Turkish   **D:** Indian

92

Which verb means 'to tap the fingers repetitively'?

A: Viola   **B:** Piano

C: Flute   **D:** Drum

93

Who is the female leader of a marching band?

A: Drum generalette   **B:** Drum colonelette

C: Drum majorette   **D:** Drum captainette

94

What is the name of the shelter beside
a sports field, where the coach sits during the game?

A: Buyout   **B:** Fallout

C: Knockout   **D:** Dugout

95

The White Cliffs of Dover mainly consist of which substance?

A: Basalt   **B:** Chalk

C: Granite   **D:** Marble

If you would like to use your 50:50 please turn to page 260
To ask the audience please turn to page 280
Turn to the answer section on page 300 to find out if you've won £200!

2♦£200

96

On which of these items would you be most likely to find a 'latch'?

A: Door

B: Saucepan

C: Light bulb

D: Bed

97

Which term of endearment is also a word meaning 'expensive'?

A: Dear

B: Darling

C: Honey

D: Beloved

98

Which family pet is most likely to be kept in a hutch?

A: Pony

B: Rabbit

C: Parrot

D: Terrapin

99

According to the saying, you should not put all your eggs into one . . . what?

A: Saucepan

B: Basket

C: Cake

D: Omelette

100

Which of these would be worn by someone trying to give up smoking?

A: Eye patch

B: Nicotine patch

C: Elbow patch

D: Cabbage patch

If you would like to use your 50:50 please turn to page 260
To ask the audience please turn to page 280
Turn to the answer section on page 300 to find out if you've won £200!

3 ♦ £300

1

Which is a contemptuous term for an environmental campaigner?

- **A:** Tree surgeon
- **B:** Tree hugger
- **C:** Tree creeper
- **D:** Tree hopper

2

Which bear has a friend called Bill Badger?

- **A:** Paddington
- **B:** Rupert
- **C:** Winnie-the-Pooh
- **D:** Baloo

3

What name is given to the track used by athletes in the long jump and triple jump?

- **A:** Piste
- **B:** Runway
- **C:** Path
- **D:** Slipway

4

Which of these is recognised as the personification of death?

- **A:** Grim Gardener
- **B:** Grim Weeder
- **C:** Grim Sower
- **D:** Grim Reaper

5

Which of these singers is not British by birth?

- **A:** Tom Jones
- **B:** Bruce Springsteen
- **C:** Charlotte Church
- **D:** Alison Moyet

If you would like to use your 50:50 please turn to page 260
To ask the audience please turn to page 280
Turn to the answer section on page 300 to find out if you've won £300!

6

What normally accompanies a coat of arms in heraldry?

A: Lotto

B: Blotto

C: Motto

D: Giotto

7

In the pantomime, Cinderella loses a slipper made of what at the ball?

A: Iron

B: Sugar

C: Glass

D: Gold

8

The river Thames flows through which English city?

A: Birmingham

B: Liverpool

C: Manchester

D: London

9

Which of these is a Canadian province?

A: Montana

B: Minnesota

C: Manitoba

D: Missouri

10

Which of these is a Gilbert and Sullivan operetta based in Japan?

A: The Subaru

B: The Mikado

C: The Mitsubishi

D: The Shogun

If you would like to use your 50:50 please turn to page 260
To ask the audience please turn to page 280
Turn to the answer section on page 300 to find out if you've won £300!

3 ◆ £300

11

Which animals are traditionally fought by matadors?

A: Lions　　　　　　　**B:** Bulls

C: Bears　　　　　　　**D:** Sharks

12

Which historic London building is locked every night in the 700-year-old Ceremony of the Keys?

A: St Paul's Cathedral　　　**B:** Buckingham Palace

C: The Tower of London　　**D:** Westminster Abbey

13

What is the traditional name for Cinderella's father in the pantomime?

A: Baron Holdup　　　　**B:** Baron Situp

C: Baron Brushup　　　　**D:** Baron Hardup

14

Which race is held over 200 laps on a 2½-mile circuit?

A: Le Mans　　　　　　**B:** Indy 500

C: Monte Carlo rally　　　**D:** Tour de France

15

What are you said to 'keep from the door' if you ward off poverty?

A: The fox　　　　　　　**B:** The tiger

C: The wolf　　　　　　　**D:** The panther

If you would like to use your 50:50 please turn to page 260
To ask the audience please turn to page 280
Turn to the answer section on page 300 to find out if you've won £300!

16

What are manufactured by the US-based company Hasbro?

A: Fragrances

B: Toys

C: Computers

D: Cars

17

In films, who replaces an actor for any dangerous scenes?

A: Dangerman

B: Riskman

C: Dareman

D: Stuntman

18

Granny Smith is a variety of which fruit?

A: Pear

B: Apple

C: Grape

D: Banana

19

Which of these is a form of hockey?

A: Splinty

B: Stinty

C: Swinty

D: Shinty

20

Which of these is a small apartment with just one main room?

A: Studio flat

B: Radio flat

C: Audio flat

D: Rodeo flat

If you would like to use your 50:50 please turn to page 260
To ask the audience please turn to page 280
Turn to the answer section on page 300 to find out if you've won £300!

21

The Pope is the Roman Catholic bishop of which city?

A: New York

B: London

C: Paris

D: Rome

22

Complete the title of the Arctic Monkey's hit single, 'I Bet You Look Good On The . . .'?

A: Dancefloor

B: Beach

C: TV

D: Tennis court

23

Which name is used to refer to a male cat?

A: Jim

B: Sam

C: Ron

D: Tom

24

Ogen, Galia and Honeydew are varieties of which fruit?

A: Tomato

B: Melon

C: Apple

D: Pear

25

In hospital, 'local' is normally an abbreviation for what?

A: Local anaesthetic

B: Local stitches

C: Local scan

D: Local therapy

If you would like to use your 50:50 please turn to page 260
To ask the audience please turn to page 280
Turn to the answer section on page 300 to find out if you've won £300!

3♦£300

26

In volleyball, the ball is kept in the air using which parts of the body?

A: Elbows

B: Hands

C: Feet

D: Knees

27

Which of these is a type of boiled sweet?

A: Barley sugar

B: Corn sugar

C: Rye sugar

D: Wheat sugar

28

In horse racing, what does the abbreviation 'SP' stand for?

A: Saddle pains

B: Starting price

C: Six ponies

D: Stallion pedigree

29

Which of these professionals should have a good 'bedside manner'?

A: Accountant

B: Barrister

C: Doctor

D: Teacher

30

In which direction does the sun rise?

A: North

B: South

C: East

D: West

If you would like to use your 50:50 please turn to pages 260 and 261
To ask the audience please turn to pages 280 and 281
Turn to the answer section on page 300 to find out if you've won £300!

3 ◆ £300

31

Which of these is a pancake-based dessert?

A: Crêpe Yvette

B: Crêpe Annette

C: Crêpe Lisette

D: Crêpe Suzette

32

Which of these is a piece of equipment for cutting paper?

A: Guillemot

B: Guinevere

C: Guillotine

D: Guinness

33

If you deal with a challenge well, you are said to 'pass with flying . . .' what?

A: Foxes

B: Pickets

C: Jackets

D: Colours

34

What is the more popular name for a meteor?

A: Shooting planet

B: Shooting sun

C: Shooting moon

D: Shooting star

35

Someone with no secrets is often described as an 'open . . .' what?

A: Mouth

B: Door

C: Hand

D: Book

If you would like to use your 50:50 please turn to page 261
To ask the audience please turn to page 281
Turn to the answer section on page 300 to find out if you've won £300!

36

What would you be most
likely to buy in a delicatessen?

A: Children's clothes

B: Electrical goods

C: Food

D: Furniture

37

Traditionally, who would be most likely
to use a plough in his work?

A: Blacksmith

B: Archer

C: Farmer

D: Greengrocer

38

What name is given to the summer period
when newspapers print very little serious news?

A: Dull days

B: Mad month

C: Silly season

D: Slow week

39

Which part of a house is most likely to be double-glazed?

A: Chimney

B: Fireplace

C: Roof

D: Windows

40

Who did King Edward VI name as his successor?

A: Lady Jane Grey

B: Lady Jane White

C: Lady Jane Black

D: Lady Jane Brown

If you would like to use your 50:50 please turn to page 261
To ask the audience please turn to page 281
Turn to the answer section on page 300 to find out if you've won £300!

3 ◆ £300

41

Which type of dog shares its name with someone who compiles crosswords?

A: Spaniel

B: Retriever

C: Terrier

D: Setter

42

If you are aware of the latest news and developments, where are you are said to have 'a finger'?

A: In the pie

B: On the pulse

C: On the nail

D: In the eye

43

Great Britain lies in which ocean?

A: Atlantic

B: Pacific

C: Arctic

D: Indian

44

What kind of animal is affectionately known as a 'pooch'?

A: Cat

B: Dog

C: Tortoise

D: Rabbit

45

What are the radiating bars of a wheel called?

A: Spleens

B: Splints

C: Spices

D: Spokes

If you would like to use your 50:50 please turn to page 261

To ask the audience please turn to page 281

Turn to the answer section on page 300 to find out if you've won £300!

3 ◆ £300

Which of these is a healer who manipulates bones?

A: Psychopath

B: Osteopath

C: Telepath

D: Gardenpath

47

Which of these is a traditional punishment at sea?

A: Walking the water

B: Walking the plank

C: Walking the tightrope

D: Walking the dog

48

Which name for a pair of trousers
can also mean 'gasps'?

A: Bags

B: Cords

C: Pants

D: Jeans

49

Which shade of pink is named after a fish?

A: Stickleback

B: Salmon

C: Shark

D: Stingray

50

In motor racing, what name is given
to the position at the front of the grid?

A: Rod position

B: Stick position

C: Pole position

D: Cane position

If you would like to use your 50:50 please turn to page 261
To ask the audience please turn to page 281
Turn to the answer section on page 300 to find out if you've won £300!

3 ♦ £300

Which of these is a type of drinking-glass?

A: Acrobat

B: Clown

C: Ringmaster

D: Tumbler

52

What does a cobbler mend for a living?

A: Televisions

B: Shoes

C: Bones

D: Computers

53

What is the name for the hard skin at the base of the fingernail?

A: Dearicle

B: Popsicle

C: Sweeticle

D: Cuticle

54

Which of these phrases means 'having nothing in common'?

A: Germans alone

B: Belgians aloof

C: Poles apart

D: French afar

55

Which of these is a specific region in north-west England?

A: River Area

B: Lake District

C: Stream Sector

D: Waterfall Domain

If you would like to use your 50:50 please turn to page 261
To ask the audience please turn to page 281
Turn to the answer section on page 300 to find out if you've won £300!

3 ◆ £300

56

Which of these is traditionally heard before a race?

- **A:** Starting pistol
- **B:** Opening gun
- **C:** Initiating rifle
- **D:** Beginning musket

57

Which of these is a timepiece?

- **A:** Sleeper clock
- **B:** Carriage clock
- **C:** Buffer clock
- **D:** Rail clock

58

What name is given to someone who spoils everyone else's fun?

- **A:** Slaughterbliss
- **B:** Killjoy
- **C:** Murdercheer
- **D:** Slaypleasure

59

Which of these is a type of straw hat?

- **A:** Boater
- **B:** Planer
- **C:** Motorer
- **D:** Trainer

60

What does a laundress do for a living?

- **A:** Looks after children
- **B:** Bakes cakes
- **C:** Washes clothes
- **D:** Grows flowers

If you would like to use your 50:50 please turn to page 261
To ask the audience please turn to page 281
Turn to the answer section on page 300 to find out if you've won £300!

61

Which of these is a style of jazz?

A: Scraptime

B: Ragtime

C: Shredtime

D: Droptime

62

Which part of the world has a name meaning 'many islands'?

A: Louisiana

B: Polynesia

C: Iberia

D: Africa

63

Someone with a low opinion of him or herself
is said to have what kind of complex?

A: Minority

B: Capability

C: Inferiority

D: Legibility

64

Which of these words refers to someone knowledgeable
or enthusiastic about a subject?

A: Buff

B: Cuff

C: Duff

D: Ruff

65

The phrase 'all the better to see you with' is associated
with which children's story?

A: Snow White

B: Red Riding Hood

C: Cinderella

D: Sleeping Beauty

If you would like to use your 50:50 please turn to page 261
To ask the audience please turn to page 281
Turn to the answer section on page 300 to find out if you've won £300!

3 ◆ £300

66

'Top of the morning' is a traditional greeting associated with which country?

- **A:** New Zealand
- **B:** Ireland
- **C:** Canada
- **D:** Wales

67

An insole is used as a lining for what?

- **A:** Briefcase
- **B:** Shopping bag
- **C:** Cutlery drawer
- **D:** Shoe

68

In relation to the Beatles, what was situated at Abbey Road in London?

- **A:** John's birthplace
- **B:** Yoko Ono's art gallery
- **C:** Band's recording studio
- **D:** Ringo's mansion

69

What is the name for the person employed to reveal the numbers in a bingo hall?

- **A:** Yeller
- **B:** Shouter
- **C:** Caller
- **D:** Screecher

70

Which word goes in front of 'reaction', 'gang' and 'letter' to make well-known phrases?

- **A:** Cable
- **B:** Chain
- **C:** Clamp
- **D:** Clasp

If you would like to use your 50:50 please turn to page 261
To ask the audience please turn to page 281
Turn to the answer section on page 300 to find out if you've won £300!

3 ♦ £300

71

What name is given to the questioning
of a witness in a court of law?

- **A:** Crossfire
- **B:** Cross-country
- **C:** Cross-examination
- **D:** Cross-pollination

72

What kind of sporting event is often known as a 'bout'?

- **A:** Tennis match
- **B:** Football match
- **C:** Boxing match
- **D:** Cricket match

73

Which word refers specifically to the removal
of wool from a sheep?

- **A:** Trimming
- **B:** Shearing
- **C:** Carving
- **D:** Whittling

74

Which part of the tea plant
is used to make the hot drink?

- **A:** Oil
- **B:** Bark
- **C:** Root
- **D:** Leaf

75

What is the name for the wrinkles
at the corner of someone's eyes?

- **A:** Crow's eggs
- **B:** Crow's feathers
- **C:** Crow's beaks
- **D:** Crow's feet

If you would like to use your 50:50 please turn to page 261
To ask the audience please turn to page 281
Turn to the answer section on page 300 to find out if you've won £300!

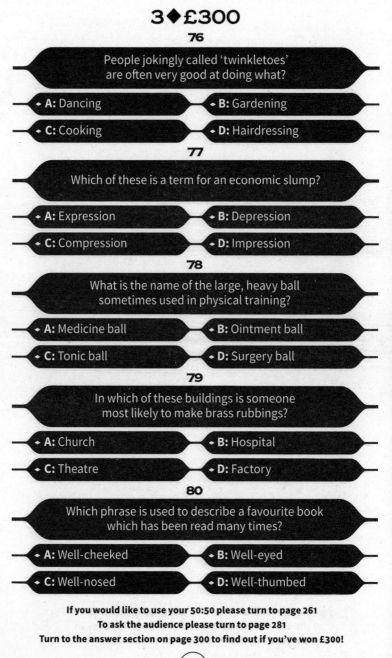

3 ◆ £300

76
People jokingly called 'twinkletoes' are often very good at doing what?

A: Dancing

B: Gardening

C: Cooking

D: Hairdressing

77
Which of these is a term for an economic slump?

A: Expression

B: Depression

C: Compression

D: Impression

78
What is the name of the large, heavy ball sometimes used in physical training?

A: Medicine ball

B: Ointment ball

C: Tonic ball

D: Surgery ball

79
In which of these buildings is someone most likely to make brass rubbings?

A: Church

B: Hospital

C: Theatre

D: Factory

80
Which phrase is used to describe a favourite book which has been read many times?

A: Well-cheeked

B: Well-eyed

C: Well-nosed

D: Well-thumbed

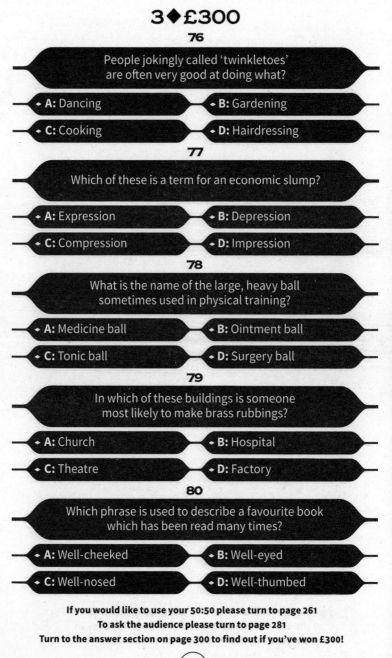

If you would like to use your 50:50 please turn to page 261
To ask the audience please turn to page 281
Turn to the answer section on page 300 to find out if you've won £300!

3 ◆ £300

Which of these is a bird of prey?

A: Vole

B: Vixen

C: Vulture

D: Viper

82

Which of these is traditionally baked in America?

A: Hop-it wasp crumble

B: Get-off bee tart

C: Shoo-fly pie

D: Begone-bug cake

83

According to the well-known saying, what comes 'before a fall'?

A: Kerb

B: Cliff

C: Pride

D: Trip

84

Which of these is the real name of a garish vivid colour?

A: Screaming yellow

B: Surprising green

C: Stunning red

D: Shocking pink

85

Which phrase refers to a conventional postal system?

A: Snail mail

B: Slug dispatch

C: Tortoise post

D: Turtle fax

If you would like to use your 50:50 please turn to pages 261 and 262
To ask the audience please turn to pages 281 and 282
Turn to the answer section on page 300 to find out if you've won £300!

3 ◆ £300

86

If 'the smoke' refers to a big town or city, which term refers to the more remote rural areas?

A: The sticks

B: The stocks

C: The stacks

D: The shacks

87

Which of these is a synonym for 'dusk'?

A: Starlight

B: Twilight

C: Moonlight

D: Sunlight

88

Which of these is not a name that identifies November 5th?

A: Firework Night

B: Guy Fawkes' Night

C: Mischief Night

D: Bonfire Night

89

Which exclamation means that a disastrous ending has occurred?

A: Nets!

B: Drapes!

C: Blinds!

D: Curtains!

90

Which of these words means to foil a plan or prevent something being accomplished?

A: Thwack

B: Thwell

C: Thwart

D: Thwaite

If you would like to use your 50:50 please turn to page 262
To ask the audience please turn to page 282
Turn to the answer section on page 300 to find out if you've won £300!

3 ♦ £300

91

What is a name for a person who hoards things of little use?

A: Goose

B: Magpie

C: Hawk

D: Gull

92

Which nursery rhyme ends with two sneezes and a fall?

A: Ring-a-ring o' Roses

B: The Mulberry Bush

C: The Big Ship Sails

D: Oranges and Lemons

93

Which of these is a narrow piece of land projecting into the sea?

A: Coat

B: Cloak

C: Cape

D: Caftan

94

At which of these battles did English longbowmen famously defeat the French?

A: Hastings

B: Rorke's Drift

C: Crécy

D: Balaclava

95

Strobe lighting is most likely to be used regularly in which of these places?

A: Operating theatre

B: Garden centre

C: Discotheque

D: Register office

If you would like to use your 50:50 please turn to page 262
To ask the audience please turn to page 282
Turn to the answer section on page 300 to find out if you've won £300!

96

How is bread without butter
or a spread on top described?

A: Arid bread

B: Dry bread

C: Desiccated bread

D: Parched bread

97

Complete the title of the George Formby song:
'When I'm Cleaning . . .'?

A: Carpets

B: Windows

C: Chimneys

D: Cupboards

98

With which element is the description
'sterling' most associated?

A: Strontium

B: Silver

C: Sodium

D: Silicon

99

Complete the phrase: 'Ignorance is . . .'?

A: Rapture

B: Delight

C: Pleasure

D: Bliss

100

When something is completely finished, it is often
described as 'done and . . .' what?

A: Driven

B: Dropped

C: Dusted

D: Docile

If you would like to use your 50:50 please turn to page 262
To ask the audience please turn to page 282
Turn to the answer section on page 300 to find out if you've won £300!

4 ◆ £500

1

Which character was played by David Schwimmer in the TV sitcom 'Friends'?

A: Joey

B: Chandler

C: Ross

D: Gunther

2

'Roll-on roll-off' is a type of which form of transport?

A: Ferry

B: Train

C: Motorcycle

D: Helicopter

3

In Britain, Father's Day falls in which month?

A: December

B: October

C: August

D: June

4

Which is not one of Winnie-the-Pooh's friends?

A: Tigger

B: Piglet

C: Eeyore

D: Monkey

5

Gordon and Irish are breeds of which dog?

A: Retriever

B: Setter

C: Poodle

D: Spaniel

If you would like to use your 50:50 please turn to page 262
To ask the audience please turn to page 282
Turn to the answer section on page 301 to find out if you've won £300!

4◆£500

6

Which of these is an informal word for 'father'?

A: Whip

B: Snap

C: Crackle

D: Pop

7

In Greek mythology, where do the gods live?

A: Delphi

B: Acropolis

C: Mount Olympus

D: Ithaca

8

What alternative name is given to Tchaikovsky's 6th Symphony?

A: Pathétique

B: Miserable

C: Distressé

D: Hôpe-lesse

9

Which teeth are usually known as wisdom teeth?

A: Canines

B: Incisors

C: Molars

D: Milk

10

The Greek god Pan is depicted with the legs of which creature?

A: Badger

B: Goat

C: Frog

D: Pig

If you would like to use your 50:50 please turn to page 262
To ask the audience please turn to page 282
Turn to the answer section on page 301 to find out if you've won £500!

11

What does it mean when food is described as 'picante'?

A: Creamy

B: Spicy

C: Slow-cooked

D: Raw

12

When two people perform an eskimo kiss, which parts of their bodies do they rub together?

A: Noses

B: Ears

C: Knees

D: Hands

13

What is the collective name for veins and arteries?

A: Blood channels

B: Blood lines

C: Blood routes

D: Blood vessels

14

Where did Audrey Hepburn have 'Breakfast at' in the title of a 1961 film?

A: Harrods

B: Claridge's

C: Selfridges

D: Tiffany's

15

What is the specific name for a book in which a stamp collection is kept?

A: Atlas

B: Folder

C: Album

D: Binder

If you would like to use your 50:50 please turn to page 262
To ask the audience please turn to page 282
Turn to the answer section on page 301 to find out if you've won £500!

4♦£500

16

What is the traditional drink on Burns Night?

- **A:** Gin
- **B:** Whisky
- **C:** Rum
- **D:** Vodka

17

George Michael was the lead singer in which musical duo?

- **A:** Damn!
- **B:** Cram!
- **C:** Wham!
- **D:** Bam!

18

Which artist is well known for cutting off his ear?

- **A:** Picasso
- **B:** Goya
- **C:** Raphael
- **D:** Van Gogh

19

In a sporting context, which colour is associated with Oxford University?

- **A:** Light blue
- **B:** Pale pink
- **C:** Dark blue
- **D:** Ruby red

20

What type of life form are mushrooms and toadstools?

- **A:** Jealousguy
- **B:** Tallguy
- **C:** Fungi
- **D:** Sadguy

If you would like to use your 50:50 please turn to page 262
To ask the audience please turn to page 282
Turn to the answer section on page 301 to find out if you've won £500!

21

Who wrote the popular 1982 animated tale 'The Snowman' about a young boy on Christmas Eve?

A: Raymond Chandler

B: Raymond Blanc

C: Raymond Briggs

D: Raymond Carver

22

What is the nationality of Agatha Christie's detective Hercule Poirot?

A: American

B: Russian

C: German

D: Belgian

23

Which animal particularly likes bamboo?

A: Lion

B: Giant panda

C: Rhino

D: Gorilla

24

What is the name of Basil Fawlty's wife in the TV sitcom 'Fawlty Towers'?

A: Stella

B: Stephanie

C: Sylvia

D: Sybil

25

Which of these countries has a red, white and blue flag?

A: Brazil

B: Sweden

C: France

D: China

If you would like to use your 50:50 please turn to page 262
To ask the audience please turn to page 282
Turn to the answer section on page 301 to find out if you've won £500!

4 ◆ £500

26

What is the first name of the broadcaster whose surname is Bragg?

- **A:** Jeremy
- **B:** Melvyn
- **C:** Kirsty
- **D:** Floella

27

In which film does Bruce Willis play a child psychologist treating a boy who can see dead people?

- **A:** The Seventh Seal
- **B:** The Sixth Sense
- **C:** The Fifth Wheel
- **D:** The Third Degree

28

What is the nickname of football club West Ham United?

- **A:** The Drillers
- **B:** The Wrenchers
- **C:** The Hammers
- **D:** The Spanners

29

Which of these phrases means 'take a break'?

- **A:** Take two
- **B:** Take three
- **C:** Take four
- **D:** Take five

30

Caviar is obtained from the eggs of which type of creature?

- **A:** Bird
- **B:** Mammal
- **C:** Insect
- **D:** Fish

If you would like to use your 50:50 please turn to page 262
To ask the audience please turn to page 282
Turn to the answer section on page 301 to find out if you've won £500!

4 ◆ £500

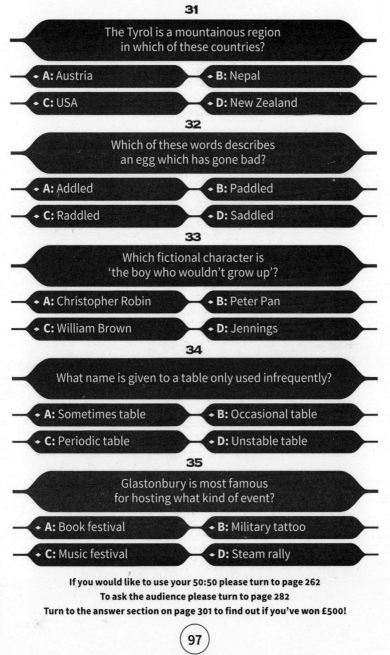

31

The Tyrol is a mountainous region
in which of these countries?

A: Austria   **B:** Nepal

C: USA   **D:** New Zealand

32

Which of these words describes
an egg which has gone bad?

A: Addled   **B:** Paddled

C: Raddled   **D:** Saddled

33

Which fictional character is
'the boy who wouldn't grow up'?

A: Christopher Robin   **B:** Peter Pan

C: William Brown   **D:** Jennings

34

What name is given to a table only used infrequently?

A: Sometimes table   **B:** Occasional table

C: Periodic table   **D:** Unstable table

35

Glastonbury is most famous
for hosting what kind of event?

A: Book festival   **B:** Military tattoo

C: Music festival   **D:** Steam rally

If you would like to use your 50:50 please turn to page 262
To ask the audience please turn to page 282
Turn to the answer section on page 301 to find out if you've won £500!

4 ◆ £500

36

Complete the famous quotation from the 1966 World Cup, 'They think it's all over . . .'?

- **A:** Another goal
- **B:** Hurrah!
- **C:** It is now
- **D:** Well, it isn't

37

Gammon comes from which animal?

- **A:** Bull
- **B:** Sheep
- **C:** Calf
- **D:** Pig

38

Which of these is a famous make of revolver?

- **A:** Mare
- **B:** Colt
- **C:** Filly
- **D:** Foal

39

Which of these is a popular garden flower?

- **A:** Nice David
- **B:** Lovely Henry
- **C:** Sweet William
- **D:** Cute Peter

40

Which anagram of 'points' refers to part of an engine?

- **A:** Notips
- **B:** Onspit
- **C:** Piston
- **D:** Spitno

If you would like to use your 50:50 please turn to page 262
To ask the audience please turn to page 282
Turn to the answer section on page 301 to find out if you've won £500!

4 ◆ £500

41

Which of these is a type of mammal?

A: Meat bat

B: Fish bat

C: Fruit bat

D: Veg bat

42

What is the surname of Sir James, the famous Irish flautist?

A: Clare

B: Galway

C: Limerick

D: Waterford

43

Which of these words describes someone with a very pale face?

A: Ashen

B: Oaken

C: Beechen

D: Pinen

44

If someone is described as 'verbose', what do they do a lot of?

A: Walking

B: Sleeping

C: Talking

D: Eating

45

Where would you be most likely to find AstroTurf?

A: Road surfaces

B: Satellite dishes

C: Frying pans

D: Sports fields

If you would like to use your 50:50 please turn to page 263
To ask the audience please turn to page 283
Turn to the answer section on page 301 to find out if you've won £500!

4♦£500

46

What is the name of the money paid when a footballer moves from one club to another?

A: Postage tax

B: Delivery cost

C: Transfer fee

D: Purchase order

47

Which of these phrases refers to something which reaches an acceptable standard?

A: Slash the salt

B: Cut the mustard

C: Break the vinegar

D: Chop the pepper

48

Which of these is an unsweetened biscuit, often eaten with cheese?

A: Cream cracker

B: Milk cracker

C: Fromage frais cracker

D: Yoghurt cracker

49

What is the slang term for a police trap set up to catch criminals?

A: Sting

B: Croak

C: Buzz

D: Bark

50

What is the name of the fairground game in which small hoops are thrown over prizes?

A: Hoop-hoop

B: Hoopy

C: Hoopla

D: Hooper

If you would like to use your 50:50 please turn to page 263
To ask the audience please turn to page 283
Turn to the answer section on page 301 to find out if you've won £500!

4 ◆ £500

51

What is the name for the small house at the gates of a park or grounds of a larger house?

- **A:** Croft
- **B:** Crypt
- **C:** Lodge
- **D:** Barn

52

Which of these is a type of warm underwear?

- **A:** Diagonal
- **B:** Thermal
- **C:** Regional
- **D:** Fraternal

53

Where on the body would you be most likely to wear pumps?

- **A:** Head
- **B:** Shoulders
- **C:** Knees
- **D:** Feet

54

In what kind of restaurant is miso soup most likely to be served?

- **A:** Indian
- **B:** Japanese
- **C:** French
- **D:** Italian

55

What is Cockney rhyming slang for the word 'wife'?

- **A:** Fork and knife
- **B:** Trouble and strife
- **C:** Drum and fife
- **D:** Death and life

If you would like to use your 50:50 please turn to page 263
To ask the audience please turn to page 283
Turn to the answer section on page 301 to find out if you've won £500!

4 ◆ £500

56

Which of these is a term for preserving meat or fish?

A: Healing

B: Mending

C: Curing

D: Convalescing

57

What kind of converter is used in a car
to help cut down on poisonous gas emissions?

A: Catastrophic

B: Cataclysmic

C: Catatonic

D: Catalytic

58

Which of these phrases refers to a
considerable sum of money?

A: Gorgeous guinea

B: Divine dollar

C: Pretty penny

D: Ravishing rouble

59

Which popular cake, made from two layers filled
with jam, shares its name with a queen of England?

A: Victoria sponge

B: Mary bun

C: Elizabeth scone

D: Anne cupcake

60

Which royal house links a type of chair with a type of knot?

A: Tudor

B: York

C: Stuart

D: Windsor

If you would like to use your 50:50 please turn to page 263
To ask the audience please turn to page 283
Turn to the answer section on page 301 to find out if you've won £500!

4 ◆ £500

61

Which of these is a term for a stretch of road on which accidents frequently occur?

A: Night spot **B:** Dark spot

C: Soft spot **D:** Black spot

62

What are you said to cross with someone if you have an argument or dispute with them?

A: Bones **B:** Eyes

C: Swords **D:** References

63

When an American says 'Check, please!' in a restaurant, what is he asking for?

A: Salt **B:** Wine

C: Ashtray **D:** Bill

64

Which of these is a fast-moving bird?

A: Streetracer **B:** Trackdasher

C: Roadrunner **D:** Lanesprinter

65

What type of theatre has seats arranged around a central stage?

A: Theatre-in-the-round **B:** Theatre-in-the-sphere

C: Theatre-in-the-circle **D:** Theatre-in-the-orb

If you would like to use your 50:50 please turn to page 263
To ask the audience please turn to page 283
Turn to the answer section on page 301 to find out if you've won £500!

4 ◆ £500

66

Which word links the sound
of laughter and the sound of bells?

- **A:** Boom
- **B:** Crash
- **C:** Peal
- **D:** Pop

67

A ship in good enough condition to
sail is said to be 'sea . . .' what?

- **A:** Able
- **B:** Creditable
- **C:** Fit
- **D:** Worthy

68

Which mushy vegetables are traditionally eaten
with fish and chips?

- **A:** Carrots
- **B:** Peas
- **C:** Sprouts
- **D:** Turnips

69

Which animals are most associated with a gymkhana?

- **A:** Sheep
- **B:** Cattle
- **C:** Dogs
- **D:** Horses

70

Which of these is famous for its caverns and stalactites?

- **A:** Stilton Valley
- **B:** Caerphilly Canyon
- **C:** Cheddar Gorge
- **D:** Roquefort Ravine

If you would like to use your 50:50 please turn to page 263
To ask the audience please turn to page 283
Turn to the answer section on page 301 to find out if you've won £500!

4♦£500

Which of these is a popular garden flower?

A: Candlemas daisy

B: Lammas daisy

C: Michaelmas daisy

D: Martinmas daisy

72

What would you be most likely to put in your corsage?

A: Wine

B: Flower

C: Rabbit

D: Shoe

73

What kind of roll are you most likely to eat in a Chinese restaurant?

A: Spring roll

B: Summer roll

C: Autumn roll

D: Winter roll

74

Which name is the rhyming slang for 'tea'?

A: Carrie Gee

B: Josie Mee

C: Annie Dee

D: Rosie Lee

75

What name is given to a type of polythene wrapping containing many small air pockets?

A: Fizz wrap

B: Pop wrap

C: Crackle wrap

D: Bubble wrap

If you would like to use your 50:50 please turn to page 263
To ask the audience please turn to page 283
Turn to the answer section on page 301 to find out if you've won £500!

4◆£500

76

What is the underground home of a rabbit called?

A: Burrow

B: Drey

C: Sett

D: Lair

77

Which of these is a move in gymnastics?

A: Handgrip

B: Handbrake

C: Handrail

D: Handspring

78

In which country was singer Bob Geldof born?

A: Wales

B: Republic of Ireland

C: Scotland

D: England

79

What goes with 'animal' and 'vegetable' to give the name of a traditional guessing game?

A: Mixture

B: Metal

C: Man

D: Mineral

80

What word describes a person who illegally gains access to private computer systems?

A: Surfer

B: Hacker

C: Looker

D: Tracker

If you would like to use your 50:50 please turn to page 263
To ask the audience please turn to page 283
Turn to the answer section on page 301 to find out if you've won £500!

4♦£500

Which football team has the
nickname 'The Red Devils'?

A: Liverpool

B: Manchester Utd

C: Sunderland

D: Arsenal

Which word refers to land which
is suitable for growing crops?

A: Arable

B: Parable

C: Risible

D: Visible

What kind of person would be
most interested in a 'black hole'?

A: Astronomer

B: Philosopher

C: Gardener

D: Mountaineer

Don King is best known as a promoter of which sport?

A: American football

B: Boxing

C: Motor racing

D: Snooker

What is heated to make caramel?

A: Egg

B: Flour

C: Salt

D: Sugar

If you would like to use your 50:50 please turn to page 263
To ask the audience please turn to page 283
Turn to the answer section on page 301 to find out if you've won £500!

4 ◆ £500

Which surname follows 'gun' and 'lock' to give
the names of two craftsmen?

A: Carter

B: Wright

C: Smith

D: Johnson

Which of these is a strand of yarn or rope?

A: Fry

B: Ply

C: Sly

D: Wry

The Cheyenne people are native to which continent?

A: North America

B: Asia

C: Africa

D: Australia

Which of these is a dance similar to the samba?

A: Super nova

B: Bossa nova

C: Casa nova

D: Terra nova

A 'tremor' is a small . . . what?

A: Avalanche

B: Tornado

C: Tidal wave

D: Earthquake

If you would like to use your 50:50 please turn to page 263
To ask the audience please turn to page 283
Turn to the answer section on page 301 to find out if you've won £500!

4 ◆ £500

Complete this Shakespeare quote:
'To be, or not to be: that is the . . .'?

A: Key

B: Problem

C: Question

D: Matter

92

Which of these is a shade of brown?

A: Guff

B: Buff

C: Ruff

D: Tuff

93

What kind of company was founded by
Thomas Cook in the 1840s?

A: Furniture

B: Travel

C: Wine

D: Publishing

94

Which of these countries is not part of Latin America?

A: Brazil

B: Canada

C: Cuba

D: Mexico

95

What is the name of a reading desk in a church?

A: Cassock

B: Vestry

C: Lectern

D: Font

If you would like to use your 50:50 please turn to page 263
To ask the audience please turn to page 283
Turn to the answer section on page 301 to find out if you've won £500!

4♦£500

Of which US city was Rudolph Giuliani the mayor?

A: Denver

B: New York

C: Baltimore

D: Saint Louis

Which character has been played on film by both Charles Laughton and Anthony Hopkins?

A: Sherlock Holmes

B: Robin Hood

C: Doctor Crippen

D: Captain Bligh

Edith Bowman is best known as what?

A: Opera singer

B: Impressionist

C: Comedian

D: Disc jockey

Which rap artist released 'The Slim Shady LP'

A: LL Cool J

B: Shaggy

C: Dr Dre

D: Eminem

What is the title of Dan Brown's best-selling book?

A: The Michelangelo Riddle

B: The Raphael Conundrum

C: The Botticelli Cipher

D: The Da Vinci Code

If you would like to use your 50:50 please turn to pages 263 and 264
To ask the audience please turn to pages 283 and 284
Turn to the answer section on page 301 to find out if you've won £500!

5 ◆ £1,000

1

Which Disney film is set in China?

- **A:** The Jungle Book
- **B:** Mulan
- **C:** Pinocchio
- **D:** The Aristocats

2

What was the original name of Sir Francis Drake's ship?

- **A:** Puffin
- **B:** Pigeon
- **C:** Plover
- **D:** Pelican

3

Who discovered penicillin in 1928?

- **A:** Joseph Lister
- **B:** James Lind
- **C:** William Harvey
- **D:** Alexander Fleming

4

The shrub Leylandii is most often used for what?

- **A:** Roofing
- **B:** Hedges
- **C:** Hats
- **D:** Fodder

5

Which punctuation mark is written as a small horizontal line?

- **A:** Semicolon
- **B:** Inverted comma
- **C:** Hyphen
- **D:** Bracket

If you would like to use your 50:50 please turn to page 264
To ask the audience please turn to page 284
Turn to the answer section on page 301 to find out if you've won £500!

5 ◆ £1,000

6

At which famous sports venue are the
Edrich Stand and Compton Stand?

A: Centre Court, Wimbledon **B:** Lord's

C: Silverstone **D:** Old Trafford

7

Which poisonous gas has the formula CO?

A: Cobalt tetroxide **B:** Chlorine oxide

C: Carbon monoxide **D:** Calcium trioxide

8

Which of these words mean a street entertainer?

A: Bluster **B:** Bolster

C: Boulder **D:** Busker

9

Prime Minister Edward Heath took part
in which sailing event in 1971?

A: Admiral's Cup **B:** Jules Verne Trophy

C: America's Cup **D:** Tall Ships Challenge

10

With which sporting event is
Jonathan Edwards most associated?

A: 100 metres **B:** Triple jump

C: Marathon **D:** Javelin

**If you would like to use your 50:50 please turn to page 264
To ask the audience please turn to page 284
Turn to the answer section on page 301 to find out if you've won £1,000!**

5 ◆ £1,000

11

One of the world's best-known operas is George Bizet's what?

A: Caramel
B: Cameron
C: Carmel
D: Carmen

12

The newspaper 'Le Monde' is published in which country?

A: France
B: Germany
C: Italy
D: Netherlands

13

At which part of a meal is a Mississippi mud pie most likely to be eaten?

A: Starter
B: Fish course
C: Main course
D: Dessert

14

In which sport might a competitor adjust a moveable wheel called the fulcrum to alter the amount of spring they receive?

A: Gymnastics
B: Show jumping
C: Tennis
D: Diving

15

Which piece of equipment is essential for competitors in a triathlon?

A: Horse
B: Bicycle
C: Canoe
D: Surfboard

If you would like to use your 50:50 please turn to page 264
To ask the audience please turn to page 284
Turn to the answer section on page 301 to find out if you've won £1,000!

5◆£1,000

16

A ship's sails are usually made from which fabric?

- **A:** Canvas
- **B:** Felt
- **C:** Hessian
- **D:** Cashmere

17

Which of these words is used to describe the transmission of audio files from the internet?

- **A:** Peacast
- **B:** Pipcast
- **C:** Podcast
- **D:** Pearcast

18

What are sometimes worn instead of spectacles?

- **A:** Contact lenses
- **B:** Goggles
- **C:** Face mask
- **D:** Visor

19

Which geographical feature is named after a letter of the Greek alphabet?

- **A:** Esker
- **B:** Kame
- **C:** Delta
- **D:** Mesa

20

What kind of domestic animal is a pug?

- **A:** Cat
- **B:** Rabbit
- **C:** Dog
- **D:** Fish

If you would like to use your 50:50 please turn to page 264
To ask the audience please turn to page 284
Turn to the answer section on page 301 to find out if you've won £1,000!

21

What name is given to the various substances
which trigger allergic reactions?

- **A:** Allegros
- **B:** Allergens
- **C:** Allegations
- **D:** Allegories

22

Greenland is considered to be part of which continent?

- **A:** Asia
- **B:** Europe
- **C:** North America
- **D:** South America

23

Robert Mugabe became president
of which country in 1987?

- **A:** South Africa
- **B:** Jamaica
- **C:** Zimbabwe
- **D:** New Zealand

24

In which country is the famous shrine at Lourdes?

- **A:** Belgium
- **B:** Luxembourg
- **C:** France
- **D:** Switzerland

25

In which of these is gravy usually served?

- **A:** Train
- **B:** Bus
- **C:** Plane
- **D:** Boat

If you would like to use your 50:50 please turn to page 264
To ask the audience please turn to page 284
Turn to the answer section on page 301 to find out if you've won £1,000!

5 ◆ £1,000

26

What does 'Mardi Gras' literally mean?

A: Wild party

B: Hello boys

C: Big parade

D: Fat Tuesday

27

Which is the short sequence of letters and numbers at the end of a postal address?

A: Barcode

B: Lettercode

C: Postcode

D: Mailcode

28

What would a Scotsman do with a bannock?

A: Play it

B: Wear it

C: Eat it

D: Toss it

29

The Saatchi brothers are most associated with which industry?

A: Advertising

B: Interior design

C: Fashion

D: Shipbuilding

30

What colour is the background of the Olympic flag?

A: Blue

B: White

C: Green

D: Gold

If you would like to use your 50:50 please turn to page 264
To ask the audience please turn to page 284
Turn to the answer section on page 301 to find out if you've won £1,000!

5 ♦ £1,000

31

In which of these sports would you attend a basho?

- **A:** Boxing
- **B:** Sumo
- **C:** Rugby
- **D:** Stock car racing

32

In World War II, Douglas Bader was best known as what?

- **A:** Soldier
- **B:** Sailor
- **C:** Pilot
- **D:** Politician

33

In which sport would you find a 'steward's enquiry'?

- **A:** Snooker
- **B:** Horse racing
- **C:** Boxing
- **D:** Motor racing

34

The 'MMR' is a vaccine against measles, mumps and which other disease?

- **A:** Chickenpox
- **B:** German measles
- **C:** Whooping-cough
- **D:** Polio

35

Which of these materials is most likely to be described as 'patent'?

- **A:** Silk
- **B:** Rubber
- **C:** Wool
- **D:** Leather

If you would like to use your 50:50 please turn to page 264
To ask the audience please turn to page 284
Turn to the answer section on page 302 to find out if you've won £1,000!

5♦£1,000

36

In geology, what is the name
of a fracture in the earth's crust?

A: Warp

B: Fault

C: Taint

D: Scratch

37

Osiris was a god in which mythology?

A: Greek

B: Roman

C: Egyptian

D: Celtic

38

What specific name is given to each of the corridors
in the House of Commons where MPs go to vote?

A: Hall

B: Lobby

C: Cloister

D: Passage

39

Which document would be endorsed with a visa?

A: Tax return

B: Television licence

C: Passport

D: Birth certificate

40

What is the full name of the card game known as 'crib'?

A: Cribo

B: Cribbage

C: Cribellum

D: Cribwork

**If you would like to use your 50:50 please turn to page 264
To ask the audience please turn to page 284
Turn to the answer section on page 302 to find out if you've won £1,000!**

5 ◆ £1,000

41

What would you be most likely to do with a joss stick?

A: Eat it

B: Throw it

C: Burn it

D: Read it

42

Which word goes with 'kith' to make a phrase relating to family and friends?

A: Kin

B: Kitten

C: Kettle

D: Kiss

43

What is the North American name for what we call 'crisps'?

A: Chips

B: Chirps

C: Chops

D: Chicks

44

Which of these is a name for a person who tells interesting and amusing stories?

A: Raconteur

B: Connoisseur

C: Masseur

D: Entrepreneur

45

Ink-jet is a type of which piece of office equipment?

A: Telephone

B: Printer

C: Stapler

D: Modem

If you would like to use your 50:50 please turn to page 264
To ask the audience please turn to page 284
Turn to the answer section on page 302 to find out if you've won £1,000!

5 ♦ £1,000

46

Which French name is the equivalent of the English name Stephen?

- **A:** Pierre
- **B:** Jean
- **C:** Etienne
- **D:** Jacques

47

Which of these words means assisting someone in a crime?

- **A:** Arisking
- **B:** Abetting
- **C:** Agambling
- **D:** Awagering

48

What is the unit of currency in New Zealand?

- **A:** Dollar
- **B:** Guilder
- **C:** Crown
- **D:** Peso

49

The song 'A Spoonful of Sugar' features in which film?

- **A:** Oliver!
- **B:** My Fair Lady
- **C:** Chitty Chitty Bang Bang
- **D:** Mary Poppins

50

Rogan josh is a dish typically found in what kind of restaurant?

- **A:** Chinese
- **B:** Indian
- **C:** French
- **D:** Spanish

If you would like to use your 50:50 please turn to page 264
To ask the audience please turn to page 284
Turn to the answer section on page 302 to find out if you've won £1,000!

5♦£1,000

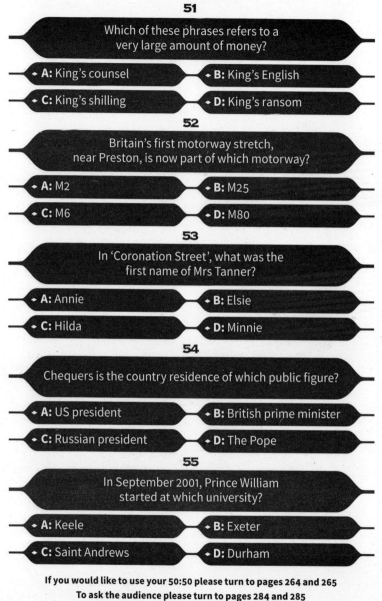

51

Which of these phrases refers to a
very large amount of money?

A: King's counsel

B: King's English

C: King's shilling

D: King's ransom

52

Britain's first motorway stretch,
near Preston, is now part of which motorway?

A: M2

B: M25

C: M6

D: M80

53

In 'Coronation Street', what was the
first name of Mrs Tanner?

A: Annie

B: Elsie

C: Hilda

D: Minnie

54

Chequers is the country residence of which public figure?

A: US president

B: British prime minister

C: Russian president

D: The Pope

55

In September 2001, Prince William
started at which university?

A: Keele

B: Exeter

C: Saint Andrews

D: Durham

If you would like to use your 50:50 please turn to pages 264 and 265
To ask the audience please turn to pages 284 and 285
Turn to the answer section on page 302 to find out if you've won £1,000!

5 ♦ £1,000

56

In Australian Slang, what is a dunny?

A: Wild dog

B: Poisonous spider

C: Dried-out creek

D: Outside toilet

57

Puff, choux and filo are all types of what?

A: Chocolate

B: Pastry

C: Batter

D: Icing

58

If you are said to 'hit it off' with someone, what do you do to them?

A: Befriend them

B: Disagree with them

C: Ignore them

D: Assault them

59

The character Penelope Pitstop featured in which television races?

A: Crazy

B: Goofy

C: Nutty

D: Wacky

60

Which of these is a type of insect?

A: Aprilbeetle

B: Mayfly

C: Junewasp

D: Julybee

If you would like to use your 50:50 please turn to page 265
To ask the audience please turn to page 285
Turn to the answer section on page 302 to find out if you've won £1,000!

5 ◆ £1,000

61

Which of these people would find perfect pitch most useful in their work?

- **A:** Vet
- **B:** Psychiatrist
- **C:** Singer
- **D:** Builder

62

What is the name for the tissue surrounding the root of a hair?

- **A:** Folderol
- **B:** Foliage
- **C:** Folio
- **D:** Follicle

63

Complete the title of the radio programme: 'Desert Island . . .'?

- **A:** Tapes
- **B:** Records
- **C:** Discs
- **D:** CDs

64

Which of these is another name for a robot?

- **A:** Automat
- **B:** Automaton
- **C:** Autobahn
- **D:** Autobiography

65

Which of these would have a spine and leaves?

- **A:** Dog
- **B:** Light bulb
- **C:** Snowflake
- **D:** Book

If you would like to use your 50:50 please turn to page 265
To ask the audience please turn to page 285
Turn to the answer section on page 302 to find out if you've won £1,000!

5 ♦ £1,000

66

What is a 'geyser'?

A: Spring

B: Valley

C: Volcano

D: Waterfall

67

The 1990 film 'Green Card' was the first English-speaking role for which French performer?

A: Yves Montand

B: Catherine Deneuve

C: Gérard Depardieu

D: Vanessa Paradis

68

In which school subject would you learn about sines and cosines?

A: Geography

B: Biology

C: Maths

D: History

69

What is the first name of Mr Gorbachev, the former Soviet leader?

A: Mikhail

B: Konstantin

C: Vladimir

D: Yuri

70

A chancel is a feature of what kind of building?

A: Warehouse

B: Airport

C: Church

D: Prison

If you would like to use your 50:50 please turn to page 265
To ask the audience please turn to page 285
Turn to the answer section on page 302 to find out if you've won £1,000!

5 ♦ £1,000

71

Which of the following is used to describe a strong, but as yet unexpressed, public opinion?

A: Whirlpool

B: Rip tide

C: Groundswell

D: Tidal wave

72

Which of these Kennedys is not a member of the American political dynasty?

A: John

B: Nigel

C: Robert

D: Edward

73

Marco Pierre White is most famous in what field?

A: Sculpture

B: Cooking

C: Fashion

D: Aviation

74

A fob is a chain usually attached to what?

A: Hat

B: Glove

C: Lamp

D: Watch

75

Which US state is closest to the Island of Cuba?

A: Alaska

B: California

C: Florida

D: Montana

If you would like to use your 50:50 please turn to page 265
To ask the audience please turn to page 285
Turn to the answer section on page 302 to find out if you've won £1,000!

5♦£1,000

76

Prosecco is a sparkling wine from which country?

A: Italy

B: Spain

C: Portugal

D: France

77

Which Shakespeare play was adapted as the hit musical 'West Side Story'?

A: The Taming of the Shrew

B: Romeo and Juliet

C: As You Like It

D: Much Ado About Nothing

78

Which dish consists of cold chicken in a curry and apricot-flavoured sauce?

A: Christening chicken

B: Celebration chicken

C: Coronation chicken

D: Ceremonial chicken

79

Where is 'colleen' a traditional name for a young woman?

A: England

B: Ireland

C: Scotland

D: Wales

80

What kind of musical instrument is a heckelphone?

A: Percussion

B: Keyboard

C: String

D: Woodwind

If you would like to use your 50:50 please turn to page 265
To ask the audience please turn to page 285
Turn to the answer section on page 302 to find out if you've won £1,000!

81

What kind of drink might be a 'single malt'?

A: Brandy

B: Gin

C: Port

D: Whisky

82

In which modern country is the site of the Battle of Agincourt?

A: France

B: Austria

C: Spain

D: Belgium

83

What is the name of the clip with long serrated 'jaws', used as a temporary connection to a battery?

A: Lion clip

B: Crocodile clip

C: Bear clip

D: Tiger clip

84

RSVP is an abbreviation for a phrase in which language?

A: Latin

B: English

C: German

D: French

85

Which French expression denotes a set meal at a fixed price?

A: À la carte

B: Table d'hôte

C: Carte blanche

D: Cordon bleu

If you would like to use your 50:50 please turn to page 265
To ask the audience please turn to page 285
Turn to the answer section on page 302 to find out if you've won £1,000!

5 ◆ £1,000

86

What is a 'magnum opus'?

A: Cannon

B: Champagne bottle

C: Great work

D: Legal document

87

What is 'mal de mer'?

A: Headache

B: Seasickness

C: Homesickness

D: Vertigo

88

Which of these playwrights is best known for his farces?

A: Harold Pinter

B: John Osborne

C: Ray Cooney

D: Arthur Miller

89

Which of these is an item of clothing?

A: Basque

B: Celt

C: Lapp

D: Slav

90

Which county is associated with a breed of bull terrier?

A: Derbyshire

B: Leicestershire

C: Staffordshire

D: Nottinghamshire

If you would like to use your 50:50 please turn to page 265
To ask the audience please turn to page 285
Turn to the answer section on page 302 to find out if you've won £1,000!

91

In which county is Land's End?

A: Kent

B: Norfolk

C: Cornwall

D: Lancashire

92

Which of these refers to a type of improvised jazz singing, in which the voice imitates instruments?

A: Scamel

B: Scat

C: Scollie

D: Scrab

93

What was the title of the long-running TV holiday programme presented by children for children?

A: 'Why Don't You . . .?'

B: 'How Can You . . .?'

C: 'Where Will You . . .?'

D: 'When Do You . . .?'

94

In total, how many letter 'D's are there in the names of the days of the week?

A: Six

B: Seven

C: Eight

D: Nine

95

How many stripes feature on the national flag of the USA?

A: Two

B: Three

C: Thirteen

D: Fifty

If you would like to use your 50:50 please turn to page 265
To ask the audience please turn to page 285
Turn to the answer section on page 302 to find out if you've won £1,000!

5 ◆ £1,000

96

What is the title earned by an outstanding chess player of the highest standard?

A: Grand Monsieur

B: Grand Maestro

C: Grand Master

D: Grand Marnier

97

Which of these is a pink dressing used on seafoods and salads?

A: Million Peak

B: Thousand Island

C: Hundred Reef

D: Dozen Bay

98

Which of these is the name of a James Bond adversary whose bowler hat has a razor-sharp brim?

A: Oddjob

B: Oddfellow

C: Oddball

D: Oddman

99

Which of these was a fashion item of the 1970s?

A: Tap jacket

B: Thermo-shirt

C: Tank top

D: Pipe pants

100

What 'is the best policy' according to the traditional proverb?

A: Honesty

B: Flattery

C: Brevity

D: Modesty

If you would like to use your 50:50 please turn to page 265
To ask the audience please turn to page 285
Turn to the answer section on page 302 to find out if you've won £1,000!

6♦£2,000

1

What, in Scotland, is a craig?

A: Shallow stream

B: Rocky hill

C: Dense wood

D: Boggy field

2

The port of Marseille is on which sea?

A: Baltic

B: Mediterranean

C: Caspian

D: Black

3

Captain Scott died in 1912 whilst on an expedition to where?

A: Amazon

B: Egypt

C: Mount Everest

D: South Pole

4

'Thought For The Day' is a feature of which
radio programme?

A: Today

B: Desert Island Discs

C: Just A Minute

D: The Archers

5

Which chemical element has the symbol Sn?

A: Selenium

B: Tungsten

C: Tin

D: Silicon

If you would like to use your 50:50 please turn to page 265
To ask the audience please turn to page 285
Turn to the answer section on page 302 to find out if you've won £2,000!

6 ◆ £2,000

6

The first recorded use of what by criminals took place in Paris in 1901?

A: Face masks

B: Guns

C: Getaway car

D: Dynamite

7

Which word specifically means an animal that feeds mainly on grass and other plants?

A: Herbivore

B: Herbaceous

C: Herbarian

D: Herborist

8

Which Dickens character always wears a wedding dress?

A: Miss Havisham

B: Little Nell

C: Betsy Trotwood

D: Jenny Wren

9

Which singer and actress formed the group Hole in 1989?

A: Courtney Love

B: Deborah Harry

C: Mariah Carey

D: Whitney Houston

10

What type of creature was a dinosaur?

A: Reptile

B: Amphibian

C: Mammal

D: Bird

If you would like to use your 50:50 please turn to page 265
To ask the audience please turn to page 285
Turn to the answer section on page 302 to find out if you've won £2,000!

6 ◆ £2,000

11

The word 'yodel' originally comes from which language?

A: English | **B:** French
C: German | **D:** Italian

12

Pearl Harbor is in which American state?

A: Alaska | **B:** California
C: Florida | **D:** Hawaii

13

The Scottish village of Alloway is famous as the birthplace of which writer?

A: Robert Burns | **B:** Charles Dickens
C: Dylan Thomas | **D:** Oscar Wilde

14

What kind of dessert is a 'gelato'?

A: Sponge cake | **B:** Creme caramel
C: Ice cream | **D:** Fruit pastry

15

Which novel was the subject of a famous 1960 trial?

A: Ulysses | **B:** Doctor Zhivago
C: Lady Chatterley's Lover | **D:** Lucky Jim

If you would like to use your 50:50 please turn to page 266
To ask the audience please turn to page 286
Turn to the answer section on page 302 to find out if you've won £2,000!

6◆£2,000

16

What kind of transport is a Cessna?

A: Aeroplane

B: Car

C: Motorcycle

D: Train

17

Which of these animals shares its name with a type of lily?

A: Lion

B: Cat

C: Mouse

D: Tiger

18

Which fictional character has a parrot called Polynesia?

A: Worzel Gummidge

B: Noddy

C: Dr Dolittle

D: Willy Wonka

19

The island of Rockall is in which ocean?

A: Atlantic

B: Pacific

C: Indian

D: Arctic

20

Which of these is a state on the east coast of the USA?

A: Maryland

B: California

C: Oregon

D: Washington

If you would like to use your 50:50 please turn to page 266
To ask the audience please turn to page 286
Turn to the answer section on page 302 to find out if you've won £2,000!

6 ◆ £2,000

21

Hell's Kitchen is an area of which US city?

A: Chicago

B: Los Angeles

C: New York

D: Boston

22

Which character links actors Peter Ustinov, Albert Finney and David Suchet?

A: Holmes

B: Poirot

C: Maigret

D: Wimsey

23

Charles Babbage's analytical engine was an early form of what?

A: Computer

B: Radio

C: Television

D: Video recorder

24

In which county is the town of Bakewell?

A: Lancashire

B: Derbyshire

C: Nottinghamshire

D: Leicestershire

25

What is the full name of the hospital commonly known as 'Bart's'?

A: Saint Bernard's

B: Saint Bridget's

C: Saint Bartholomew's

D: Saint Bernadette's

If you would like to use your 50:50 please turn to page 266
To ask the audience please turn to page 286
Turn to the answer section on page 302 to find out if you've won £2,000!

6 ♦ £2,000

26

The footballer Pelé played for which country?

- **A:** Argentina
- **B:** Brazil
- **C:** Colombia
- **D:** Ecuador

27

With which type of music is Shania Twain most associated?

- **A:** Country
- **B:** Jazz
- **C:** Opera
- **D:** Hip-hop

28

Typically, how often is an almanac published?

- **A:** Daily
- **B:** Weekly
- **C:** Monthly
- **D:** Yearly

29

A 'windfall' is unexpected . . . what?

- **A:** Illness
- **B:** Sleepiness
- **C:** Good luck
- **D:** Power cut

30

Which of these is a machine for making coffee?

- **A:** Generator
- **B:** Radiator
- **C:** Accumulator
- **D:** Percolator

If you would like to use your 50:50 please turn to page 266
To ask the audience please turn to page 286
Turn to the answer section on page 302 to find out if you've won £2,000!

6 ◆ £2,000

31

Which first name is shared by actress Collins and singer Armatrading?

A: Jean

B: Joan

C: Janice

D: Janet

32

What is the name of Andrew Lloyd Webber's cello-playing brother?

A: James

B: Julian

C: Jocelyn

D: Jeffrey

33

What nationality is the athlete Sonia O'Sullivan?

A: American

B: Canadian

C: Australian

D: Irish

34

Triads are secret societies originating in which country?

A: Zimbabwe

B: Uruguay

C: Bulgaria

D: China

35

What does the name Rio de Janeiro mean?

A: River of January

B: River of the Kings

C: River of Light

D: River of Hope

If you would like to use your 50:50 please turn to page 266
To ask the audience please turn to page 286
Turn to the answer section on page 302 to find out if you've won £2,000!

6 ◆ £2,000

36

What is the form of billiards played in pubs, in which balls are struck into holes guarded by pegs?

A: Bar billiards

B: Saloon billiards

C: Snug billiards

D: Tap billiards

37

The musicians Jarvis Cocker and Joe Cocker are both from which city?

A Newcastle

B Birmingham

C Sheffield

D Winchester

38

Who was the long-time presenter of BBC's 'Record Breakers'?

A: Bob Holness

B: Leslie Crowther

C: Ed Stewart

D: Roy Castle

39

Elton John's original 'Candle in the Wind' was a song about whom?

A: Kim Novak

B: Grace Kelly

C: Marilyn Monroe

D: Audrey Hepburn

40

Traditionally, what goes on top of a Lancashire hotpot?

A: Mozzarella

B: Pastry

C: Breadcrumbs

D: Potatoes

If you would like to use your 50:50 please turn to page 266
To ask the audience please turn to page 286
Turn to the answer section on page 302 to find out if you've won £2,000!

6 ♦ £2,000

41

Eritrea is a country on which continent?

A: Africa

B: Europe

C: South America

D: Asia

42

The University of Strathclyde is based in which city?

A: Belfast

B: Leeds

C: Glasgow

D: Swansea

43

The Bay of Bengal is an arm of which ocean?

A: Indian

B: Pacific

C: Atlantic

D: Arctic

44

Which of these is not a bird?

A: Sand martin

B: Sandpiper

C: Sand plover

D: Sand wedge

45

Which of these places has the same name as the county it is in?

A: Chelmsford

B: Durham

C: Exeter

D: Farnham

If you would like to use your 50:50 please turn to page 266
To ask the audience please turn to page 286
Turn to the answer section on page 302 to find out if you've won £2,000!

6 ◆ £2,000

46

The flowers of the herb chamomile are often used to make which sort of beverage?

A: Tea

B: Lemonade

C: Ginger ale

D: Wine

47

Which of these organisations represents the interests of professional football players?

A: PFA

B: PGA

C: PTA

D: PLA

48

What is the name for the junction where two rivers meet?

A: Affluence

B: Confluence

C: Effluence

D: Influence

49

Which name is used for 'J' in the phonetic alphabet?

A: Juliet

B: Joanna

C: John

D: Jeremy

50

What is the occupation of an attaché?

A: Diplomat

B: Solicitor

C: Teacher

D: Journalist

If you would like to use your 50:50 please turn to page 266
To ask the audience please turn to page 286
Turn to the answer section on page 302 to find out if you've won £2,000!

6 ◆ £2,000

51

Referring to the musical instrument, 'piano' is a shortened form of which word?

- **A:** Pianomezzo
- **B:** Pianoforte
- **C:** Pianocello
- **D:** Pianobasso

52

Which of these films, featuring the music of the Beatles, is mainly an animation?

- **A:** Magical Mystery Tour
- **B:** Yellow Submarine
- **C:** Help!
- **D:** A Hard Day's Night

53

The pre-decimal shilling was known by what informal name?

- **A:** Bill
- **B:** Ben
- **C:** Bob
- **D:** Bert

54

Which of these is not a traditional ingredient of a ploughman's lunch?

- **A:** Bread
- **B:** Cheese
- **C:** Ham
- **D:** Pickle

55

What kind of creature is a petrel?

- **A:** Bird
- **B:** Insect
- **C:** Mammal
- **D:** Fish

If you would like to use your 50:50 please turn to page 266
To ask the audience please turn to page 286
Turn to the answer section on page 302 to find out if you've won £2,000!

6 ◆ £2,000

56

What is the name for the initial stages of growth in a seed or pollen grain?

- **A:** Germination
- **B:** Integration
- **C:** Determination
- **D:** Extermination

57

If you take back something you said, you are said to 'eat your . . .' what?

- **A:** Letters
- **B:** Words
- **C:** Syllables
- **D:** Paragraphs

58

Where in Britain was the world's first metal bridge built?

- **A:** Copperbridge Gorge
- **B:** Ironbridge Gorge
- **C:** Leadbridge Gorge
- **D:** Tinbridge Gorge

59

Which of these is the title of a film and a novel by Martin Cruz Smith?

- **A:** Central Park
- **B:** Gorky Park
- **C:** Regent's Park
- **D:** MacArthur Park

60

In which gymnastics event can male competitors be helped onto the apparatus?

- **A:** Pommel horse
- **B:** Vault
- **C:** Rings
- **D:** Floor exercises

If you would like to use your 50:50 please turn to page 266
To ask the audience please turn to page 286
Turn to the answer section on page 302 to find out if you've won £2,000!

6◆£2,000

61

Who built the first fort on the site which later became known as the Tower of London?

A: Henry VIII

B: Richard the Lionheart

C: William of Orange

D: William the Conqueror

62

Complete the name of the football club: 'Rushden and . . .'?

A: Emeralds

B: Rubies

C: Sapphires

D: Diamonds

63

What name is given to the practice of passing off someone else's work as your own?

A: Pluralism

B: Paternalism

C: Plagiarism

D: Primitivism

64

Which of these film actors was born in Britain?

A: James Cagney

B: James Mason

C: James Coburn

D: James Caan

65

Tony McCoy and Adrian Maguire are famous names in which sport?

A: Golf

B: Darts

C: Horse racing

D: Rallying

If you would like to use your 50:50 please turn to page 266
To ask the audience please turn to page 286
Turn to the answer section on page 303 to find out if you've won £2,000!

6 ◆ £2,000

66

How many animal signs are there in the Chinese calendar?

- **A:** Three
- **B:** Twelve
- **C:** Seventeen
- **D:** Forty

67

Where is Osborne House, the former home of Queen Victoria?

- **A:** Windsor
- **B:** Norfolk
- **C:** Scotland
- **D:** Isle of Wight

68

Which of these is not one of the Benelux countries?

- **A:** France
- **B:** Belgium
- **C:** Luxembourg
- **D:** Netherlands

69

What did the Romans call a marketplace or public meeting place?

- **A:** Curia
- **B:** Basilica
- **C:** Forum
- **D:** Atrium

70

Who composed 'Rhapsody in Blue'?

- **A:** Cole Porter
- **B:** Irving Berlin
- **C:** George Gershwin
- **D:** Aaron Copland

If you would like to use your 50:50 please turn to pages 266 and 267
To ask the audience please turn to pages 286 and 287
Turn to the answer section on page 303 to find out if you've won £2,000!

6 ◆ £2,000

71

In which of these countries is Castilian most likely to be spoken?

A: Denmark

B: Greece

C: Luxembourg

D: Spain

72

Robert Plant and Jimmy Page are associated with which rock band?

A: Led Zeppelin

B: Iron Maiden

C: Hawkwind

D: Whitesnake

73

In the Old Testament, which book immediately follows Genesis?

A: Ecclesiastes

B: Deuteronomy

C: Exodus

D: Judges

74

Which of these is not a type of headwear?

A: Diadem

B: Coronet

C: Tiara

D: Locket

75

What is the name of a famous English landscape artist born in 1776?

A: Constable

B: Sergeant

C: Sheriff

D: Inspector

If you would like to use your 50:50 please turn to page 267
To ask the audience please turn to page 287
Turn to the answer section on page 303 to find out if you've won £2,000!

7 ◆ £4,000

1

Which of these famous poems is narrated to a guest on the way to a wedding feast?

- **A:** Ode to a Nightingale
- **B:** Daffodils
- **C:** The Rime of the Ancient Mariner
- **D:** The Lady of Shalott

2

Which of the following was an activity practised by medieval knights on horseback?

- **A** Jesting
- **B** Jousting
- **C** Jostling
- **D** Joisting

3

Who directed 'The Imaginarium of Doctor Parnassus'?

- **A:** Terry Gilliam
- **B:** Tim Burton
- **C:** Sam Mendes
- **D:** Peter Jackson

4

Which of these birds do not appear in the song 'The Twelve Days of Christmas'?

- **A:** Doves
- **B:** Hens
- **C:** Robins
- **D:** Swans

5

According to Wordsworth, what were 'Beside the lake, beneath the trees, Fluttering and dancing in the breeze'?

- **A:** Birds
- **B:** Fairies
- **C:** Daffodils
- **D:** Butterflies

If you would like to use your 50:50 please turn to page 267
To ask the audience please turn to page 287
Turn to the answer section on page 303 to find out if you've won £4,000!

6

Which tag line is associated with the 1996 Tom Cruise film 'Jerry Maguire'?

A: 30% is mine

B: Green is best

C: Show me the money

D: I just love rich

7

Miss Adelaide is a character from which musical?

A: Show Boat

B: South Pacific

C: Guys and Dolls

D: Carousel

8

The Battle of Trafalgar was fought during which major conflict?

A: Hundred Years' War

B: Napoleonic Wars

C: Boer War

D: Wars of the Roses

9

Cardialgia is another name for which medical condition?

A: Hayfever

B: Hiccups

C: Headache

D: Heartburn

10

Lerwick is the capital of which group of Scottish islands?

A: Shetlands

B: Orkneys

C: Outer Hebrides

D: Inner Hebrides

If you would like to use your 50:50 please turn to page 267
To ask the audience please turn to page 287
Turn to the answer section on page 303 to find out if you've won £4,000!

7 ◆ £4,000

11

What nationality was Alfred Nobel,
the founder of the Nobel prizes?

A: German

B: Belgian

C: Dutch

D: Swedish

12

In 1581, Sir Francis Drake became mayor of which city?

A: Hull

B: Glasgow

C: Plymouth

D: Bristol

13

The Rembrandt House is a museum in which European city?

A: Paris

B: Amsterdam

C: Stockholm

D: Vienna

14

Languedoc is an area of which European country?

A: France

B: Germany

C: Portugal

D: Austria

15

Which of these film personalities was not born in London?

A: Alfred Hitchcock

B: Grace Kelly

C: Elizabeth Taylor

D: Charlie Chaplin

If you would like to use your 50:50 please turn to page 267
To ask the audience please turn to page 287
Turn to the answer section on page 303 to find out if you've won £4,000!

7◆£4,000

16

Which of these is a salary or allowance
paid to a clergyman?

A Surplice   B Diocese

C Stipend   D Curate

17

The town of Telford is named after Thomas Telford,
who was famous in which field?

A: Poetry   **B:** Finance

C: Engineering   **D:** Music

18

Eton College is on the bank of which river?

A: Severn   **B:** Forth

C: Thames   **D:** Tamar

19

Who is traditionally thought to have been struck in the
eye by an arrow at the Battle of Hastings?

A: Harold   **B:** Henry

C: Hubert   **D:** Helvig

20

Which film musical is set in and around Rydell High?

A: Moulin Rouge   **B:** Grease

C: Fame   **D:** Little Shop of Horrors

If you would like to use your 50:50 please turn to page 267
To ask the audience please turn to page 287
Turn to the answer section on page 303 to find out if you've won £4,000!

7 ♦ £4,000

21

What is the main language of the Dominican Republic?

A: English

B: French

C: Spanish

D: Italian

22

Who founded the Prince's Trust in 1976 to give special help to the young unemployed?

A: Prince Philip

B: Prince Charles

C: Prince Andrew

D: Prince Edward

23

Which of these is a traditional ingredient of Scottish broth?

A: Porridge Oats

B: Rice

C: Pearl Barley

D: Rye

24

What does the 'A' stand for in the acronym for the army division SAS?

A: Army

B: Attack

C: Aquatic

D: Air

25

Which alcoholic drink is known as 'mother's ruin'?

A: Brandy

B: Cider

C: Gin

D: Stout

If you would like to use your 50:50 please turn to page 267
To ask the audience please turn to page 287
Turn to the answer section on page 303 to find out if you've won £4,000!

26

Which of these foods is most likely to contain gluten?

A: Milk | **B:** Bread

C: Coffee | **D:** Meat

27

What was the first name of the Hollywood star whose surname was de Havilland?

A: Marilyn | **B:** Audrey

C: Olivia | **D:** Rita

28

Where in Britain is the RSC based?

A: Newcastle-upon-Tyne | **B:** Ashton-under-Lyne

C: Weston-super-Mare | **D:** Stratford-upon-Avon

29

Which of these words refers to the mending of broken bones?

A: Darning | **B:** Sewing

C: Crocheting | **D:** Knitting

30

Who invented the Kodak camera?

A: Eastman | **B:** Westman

C: Northman | **D:** Southman

If you would like to use your 50:50 please turn to page 267
To ask the audience please turn to page 287
Turn to the answer section on page 303 to find out if you've won £4,000!

7 ♦ £4,000

31

Which of these is not classed as a precious metal?

A: Gold

B: Silver

C: Platinum

D: Zinc

32

A triumvirate is a group of three what?

A: Men

B: Atoms

C: Legs

D: Paintings

33

The mountain beaver is native to which continent?

A: North America

B: Asia

C: Europe

D: South America

34

Where is The Royal Mint based?

A: England

B: Northern Ireland

C: Scotland

D: Wales

35

In India, what is 'pachisi'?

A: Religious sect

B: Board game

C: Metallic fabric

D: Buffalo milk

If you would like to use your 50:50 please turn to page 267
To ask the audience please turn to page 287
Turn to the answer section on page 303 to find out if you've won £4,000!

7 ◆ £4,000

36

The Levant is an archaic term for the eastern countries of which region?

A: Scandinavia

B: Middle East

C: Mediterranean

D: Southeast Asia

37

What would you normally do with a demitasse?

A: Eat it

B: Drink it

C: Wear it

D: Ride it

38

With which sport is Flushing Meadow, New York most associated?

A: Baseball

B: American football

C: Ice hockey

D: Tennis

39

Grey and red are the two types of which fish?

A: Trout

B: Plaice

C: Eel

D: Mullet

40

Which of these words is derived from the name of the writer the Marquis de Sade?

A: Sadness

B: Saddle

C: Sadist

D: Sadler

If you would like to use your 50:50 please turn to page 267
To ask the audience please turn to page 287
Turn to the answer section on page 303 to find out if you've won £4,000!

41

The Mysterons were the enemies of which of these characters?

- **A:** Doctor Who
- **B:** Captain Scarlet
- **C:** Man from U.N.C.L.E.
- **D:** James Bond

42

In the Middle Ages, which fabric was worn as a sign of penitence?

- **A:** Silk
- **B:** Velvet
- **C:** Sackcloth
- **D:** Taffeta

43

Which of these dinosaurs had three horns?

- **A:** Tyrannosaurus
- **B:** Brontosaurus
- **C:** Triceratops
- **D:** Stegosaurus

44

Which wine shares its name with a joint of pork?

- **A:** Champagne
- **B:** Hock
- **C:** Retsina
- **D:** Claret

45

Which cookery term comes from the French word meaning 'to roll'?

- **A:** Soufflé
- **B:** Eclair
- **C:** Croissant
- **D:** Roulade

If you would like to use your 50:50 please turn to page 267
To ask the audience please turn to page 287
Turn to the answer section on page 303 to find out if you've won £4,000!

7 ◆ £4,000

46

In which country is the city of Marrakesh?

A: Egypt

B: Libya

C: Algeria

D: Morocco

47

What relation is the present Duke of Gloucester to Queen Elizabeth II?

A: Son

B: Cousin

C: Uncle

D: Grandson

48

In which part of the British Isles is Harlech Castle?

A: Wales

B: Scotland

C: England

D: Ireland

49

Where in London can Francis Chichester's boat Gipsy Moth IV be seen?

A: British Museum

B: Greenwich

C: St James's Palace

D: Hyde Park

50

The horror film 'The Shining' is based on a novel by which writer?

A: Michael Crichton

B: Raymond Chandler

C: John Grisham

D: Stephen King

If you would like to use your 50:50 please turn to pages 267 and 268
To ask the audience please turn to pages 287 and 288
Turn to the answer section on page 303 to find out if you've won £4,000!

7 ◆ £4,000

What nationality is the film star Antonio Banderas?

A: French

B: Italian

C: Spanish

D: Portuguese

Which popular cocktail contains a mixture of advocaat and lemonade?

A: Screwdriver

B: Sidecar

C: Snowball

D: Singapore sling

Which of these US cities was founded by Mormons?

A: Philadelphia

B: Salt Lake City

C: Miami

D: Dallas

Great Yarmouth is in which county?

A: North Yorkshire

B: Lancashire

C: Norfolk

D: Suffolk

In which sport did Ray Illingworth captain England?

A: Football

B: Cricket

C: Rugby union

D: Rugby league

If you would like to use your 50:50 please turn to page 268
To ask the audience please turn to page 288
Turn to the answer section on page 303 to find out if you've won £4,000!

7 ◆ £4,000

56

In which decade did John McEnroe win all his Wimbledon singles titles?

A: 1960s

B: 1970s

C: 1980s

D: 1990s

57

Which of the Gospel writers is the patron saint of Venice?

A: Matthew

B: Mark

C: Luke

D: John

58

Which of these is a town in North Yorkshire?

A: Hopton

B: Skipton

C: Leapton

D: Jumpton

59

Which word represents the last letter in the NATO alphabet?

A: Zebra

B: Zulu

C: Zip

D: Zephyr

60

Which of these means 'to turn upside down'?

A: Turn tortoise

B: Turn toucan

C: Turn turtle

D: Turn turkey

If you would like to use your 50:50 please turn to page 268
To ask the audience please turn to page 288
Turn to the answer section on page 303 to find out if you've won £4,000!

7 ◆ £4,000

61

In which country was the Honda motor company founded?

- **A:** Japan
- **B:** France
- **C:** Sweden
- **D:** Germany

62

Against whom did Queen Boudicca lead a revolt?

- **A:** Normans
- **B:** Vikings
- **C:** Scots
- **D:** Romans

63

Which Alfred Hitchcock film featured the character Norman Bates?

- **A:** Blackmail
- **B:** Rebecca
- **C:** The Birds
- **D:** Psycho

64

What kind of food was Lymeswold?

- **A:** Biscuit
- **B:** Potato
- **C:** Apple
- **D:** Cheese

65

Which of these was a presenter of 'Blue Peter'?

- **A:** Anthea Turner
- **B:** Phillip Schofield
- **C:** Emma Forbes
- **D:** Timmy Mallett

If you would like to use your 50:50 please turn to page 268
To ask the audience please turn to page 288
Turn to the answer section on page 303 to find out if you've won £4,000!

7 ♦ £4,000

Where did Florence Nightingale establish a hospital to tend the wounded of the Crimean War?

A: Alexandria

B: Darjeeling

C: Scutari

D: Constantinople

In the West End, the show 'The Play What I Wrote' was a tribute to which comedy duo?

A: Abbott and Costello

B: Laurel and Hardy

C: Flanagan and Allen

D: Morecambe and Wise

What is a sampan?

A: Fish

B: Boat

C: Shoe

D: Hat

Who wrote sonnets dedicated to an unidentified 'dark lady'?

A: Byron

B: Shelley

C: Blake

D: Shakespeare

In which country is the port of Haifa?

A: Israel

B: Libya

C: India

D: Lebanon

If you would like to use your 50:50 please turn to page 268
To ask the audience please turn to page 288
Turn to the answer section on page 303 to find out if you've won £4,000!

7 ◆ £4,000

71

Which of these 'seas' is actually a large
inland lake between Europe and Asia?

A: Caspian

B: Mediterranean

C: Arabian

D: Tasman

72

Birmingham and Mobile are both in which
of these American states?

A: Alabama

B: Missouri

C: Florida

D: Texas

73

Baroness Marie-Christine von Reibnitz
is better known by what name?

A: Marie Curie

B: Mata Hari

C: Madame de Pompadour

D: Princess Michael of Kent

74

Which jazz singer was famous for her song 'Fever'?

A: Billie Holiday

B: Nina Simone

C: Ella Fitzgerald

D: Peggy Lee

75

By what name are the musicians
Tom Rowlands and Ed Simons collectively known?

A: Chemical Mothers

B: Chemical Fathers

C: Chemical Brothers

D: Chemical Sisters

If you would like to use your 50:50 please turn to page 268
To ask the audience please turn to page 288
Turn to the answer section on page 303 to find out if you've won £4,000!

8 ♦ £8,000

1

Which of these creatures is a decapod?

A: Limpet
B: Starfish
C: Lobster
D: Scorpion

2

Which country is known as 'Bharat' in its own language?

A: India
B: Japan
C: Finland
D: Switzerland

3

How is Margaret Mary Emily Hyra more familiarly known?

A: Megan Fox
B: Maggie Smith
C: Meg Ryan
D: Margot Fonteyn

4

Which country would appear first in an alphabetical list of the member states of the EU?

A: Belgium
B: Austria
C: Albania
D: Bulgaria

5

What has 2½-foot eyes, a 3-foot mouth, and a 42-foot arm?

A: Angel of the North
B: Statue of Liberty
C: David
D: The Little Mermaid

If you would like to use your 50:50 please turn to page 268
To ask the audience please turn to page 288
Turn to the answer section on page 303 to find out if you've won £8,000!

8 ♦ £8,000

6

What is described in a famous poem as 'season of mists and mellow fruitfulness'?

A: Spring

B: Summer

C: Autumn

D: Winter

7

A misanthrope is someone who specifically dislikes what?

A: Dogs

B: Bad weather

C: Mankind

D: Noise

8

Which stadium is home to Southampton FC?

A: Riverside

B: St Mary's

C: King Power

D: Pride Park

9

Which specific term describes a lettuce which has suddenly and prematurely gone to seed?

A: Bolted

B: Welshed

C: Scuttled

D: Blown

10

The magazine of whose appreciation society is entitled 'Railway Cuttings'?

A: Sid James

B: Tony Hancock

C: Dick Emery

D: Frankie Howerd

If you would like to use your 50:50 please turn to page 268
To ask the audience please turn to page 288
Turn to the answer section on page 303 to find out if you've won £8,000!

8◆£8,000

11

By what name is the writer Charles Lutwidge Dodgson better known?

A: Captain Marryat

B: Lewis Carroll

C: C S Lewis

D: Hugh Lofting

12

In the Bible, how long did Jonah spend in the belly of the 'great fish'?

A: Three weeks

B: Forty days

C: One year

D: Three days

13

Who was David Beckham's best man when he married Victoria in 1999?

A: Gary Neville

B: Roy Keane

C: Ryan Giggs

D: Sir Alex Ferguson

14

Oakwell is the home of which Yorkshire football club?

A: Barnsley

B: Bradford

C: Huddersfield

D: Rotherham

15

What is the English translation of the Latin 'Magna Carta'?

A: Large map

B: Royal warrant

C: Grand scheme

D: Great charter

If you would like to use your 50:50 please turn to page 268
To ask the audience please turn to page 288
Turn to the answer section on page 303 to find out if you've won £8,000!

16

In Greek mythology, who was the father of Icarus?

A: Daedalus

B: Perseus

C: Chiron

D: Tantalus

17

What does a numismatist collect?

A: Stamps

B: Fossils

C: Teddy bears

D: Coins

18

Traditionally, pewter was a blend of tin and which other metal?

A: Copper

B: Lead

C: Silver

D: Iron

19

Which senior pupil bullied Tom Brown in the story 'Tom Brown's Schooldays'?

A: Flashman

B: Clough

C: Bartholomew

D: Harrowell

20

If a ship is rounding the 'Horn', in which part of the world is it sailing?

A: North America

B: South America

C: East Africa

D: South Africa

If you would like to use your 50:50 please turn to page 268
To ask the audience please turn to page 288
Turn to the answer section on page 303 to find out if you've won £8,000!

21

What is the name of Scrooge's late business partner in Dickens' 'A Christmas Carol'?

A: Isaac Farley

B: Abraham Warley

C: Moses Barley

D: Jacob Marley

22

In heraldry, if an animal is described as 'sejant' what is it doing?

A: Sitting upright

B: Sleeping

C: Rearing

D: Turning round

23

Which character in Shakespeare says 'Et tu Brute'?

A: King Lear

B: Julius Caesar

C: Henry V

D: Shylock

24

In which country is a policeman known as a 'carabiniere'?

A: Spain

B: France

C: Italy

D: Netherlands

25

In which century did the First Crusade begin?

A: 7th

B: 9th

C: 13th

D: 11th

If you would like to use your 50:50 please turn to page 268
To ask the audience please turn to page 288
Turn to the answer section on page 303 to find out if you've won £8,000!

26

'Ciao' is a greeting in which language?

◆ **A:** French   ◆ **B:** Italian

◆ **C:** Spanish   ◆ **D:** Greek

27

Bandicoots are native to which of these countries?

◆ **A:** Australia   ◆ **B:** China

◆ **C:** South Africa   ◆ **D:** United States

28

The Pilgrim Fathers are most associated with which ship?

◆ **A:** Mary Rose   ◆ **B:** Golden Hind

◆ **C:** Mayflower   ◆ **D:** Santa Maria

29

Which of these men was not one of the Goons?

◆ **A:** Benny Hill   ◆ **B:** Peter Sellers

◆ **C:** Spike Milligan   ◆ **D:** Harry Secombe

30

In which city is the famous Juilliard School situated?

◆ **A:** San Francisco   ◆ **B:** Chicago

◆ **C:** New York   ◆ **D:** Atlanta

If you would like to use your 50:50 please turn to page 268
To ask the audience please turn to page 288
Turn to the answer section on page 303 to find out if you've won £8,000!

8◆£8,000

31

Which of the following conditions is a form of bursitis?

A: Mumps

B: Housemaid's knee

C: Hiccups

D: Hay fever

32

Which word describes an animal's tail which can grip like a hand?

A: Prehensile

B: Prefigured

C: Preliminary

D: Preoccupied

33

Which of these is not an answer when a whole number is squared?

A: 36

B: 64

C: 81

D: 120

34

What is the main ingredient of the dessert pannacotta?

A: Biscuits

B: Sponge

C: Mascarpone cheese

D: Cream

35

Which of these is a Stephen King best-seller?

A: It

B: And

C: That

D: So

If you would like to use your 50:50 please turn to page 269
To ask the audience please turn to page 289
Turn to the answer section on page 304 to find out if you've won £8,000!

8◆£8,000

Which saint is often depicted with a spiked wheel?

A: Anne

B: Sebastian

C: Catherine

D: Alban

37

In Greek mythology, who wore wings of wax which melted when he flew near the sun?

A: Oedipus

B: Achilles

C: Icarus

D: Heracles

38

Which band, formed in the 1960s, took its name from a song by blues singer Muddy Waters?

A: Beatles

B: Moody Blues

C: Procol Harum

D: Rolling Stones

39

Which French phrase is used to describe something of the highest quality?

A: De trop

B: De luxe

C: De rigueur

D: De nous

40

Opium is derived from which flower?

A: Cornflower

B: Foxglove

C: Crocus

D: Poppy

If you would like to use your 50:50 please turn to page 269
To ask the audience please turn to page 289
Turn to the answer section on page 304 to find out if you've won £8,000!

8 ◆ £8,000

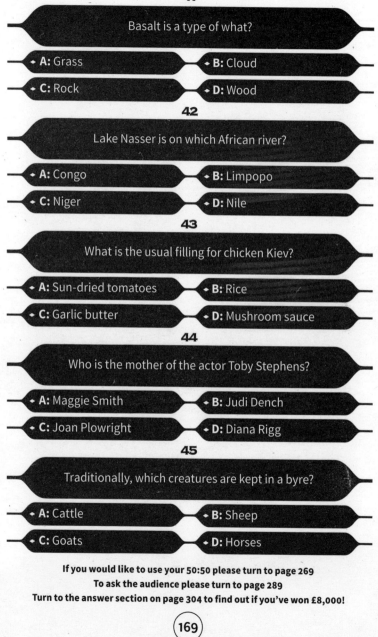

41

Basalt is a type of what?

- **A:** Grass
- **B:** Cloud
- **C:** Rock
- **D:** Wood

42

Lake Nasser is on which African river?

- **A:** Congo
- **B:** Limpopo
- **C:** Niger
- **D:** Nile

43

What is the usual filling for chicken Kiev?

- **A:** Sun-dried tomatoes
- **B:** Rice
- **C:** Garlic butter
- **D:** Mushroom sauce

44

Who is the mother of the actor Toby Stephens?

- **A:** Maggie Smith
- **B:** Judi Dench
- **C:** Joan Plowright
- **D:** Diana Rigg

45

Traditionally, which creatures are kept in a byre?

- **A:** Cattle
- **B:** Sheep
- **C:** Goats
- **D:** Horses

If you would like to use your 50:50 please turn to page 269
To ask the audience please turn to page 289
Turn to the answer section on page 304 to find out if you've won £8,000!

8 ◆ £8,000

46

Who was the third president of the USA?

A: Thomas Jefferson | **B:** John Adams
C: Abraham Lincoln | **D:** James Madison

47

Bikini Atoll lies in which ocean?

A: Atlantic | **B:** Pacific
C: Arctic | **D:** Indian

48

'Butternut' is a variety of which food?

A: Squash | **B:** Pepper
C: Onion | **D:** Potato

49

The Escorial is a palace near which capital city?

A: Moscow | **B:** Madrid
C: Vienna | **D:** Rome

50

The city of Krakow is in which country?

A: Bulgaria | **B:** Latvia
C: Poland | **D:** Russia

If you would like to use your 50:50 please turn to page 269
To ask the audience please turn to page 289
Turn to the answer section on page 304 to find out if you've won £8,000!

8♦£8,000

What is the common name for the chemical potassium nitrate?

A: Baking powder

B: Epsom salts

C: Washing soda

D: Saltpetre

In which county is the town of Chesterfield?

A: Leicestershire

B: Nottinghamshire

C: Derbyshire

D: Warwickshire

The Fabia and the Octavia are models made by which car manufacturer?

A: Toyota

B: Renault

C: Fiat

D: Skoda

Who topped the UK charts in 1972 with 'School's Out'?

A: Rod Stewart

B: Alice Cooper

C: Marc Bolan

D: Chuck Berry

The song 'The Battle Hymn of the Republic' is sung to which tune?

A: John Brown's Body

B: My Grandfather's Clock

C: Marching Through Georgia

D: Swing Low Sweet Chariot

If you would like to use your 50:50 please turn to page 269
To ask the audience please turn to page 289
Turn to the answer section on page 304 to find out if you've won £8,000!

What nationality was the explorer Vasco da Gama?

A: Italian **B:** Portuguese

C: Spanish **D:** English

57

Who belong to the trade union the NASUWT?

A: Architects **B:** Social Workers

C: Teachers **D:** Accountants

58

The designer Manolo Blahnik is most associated with which fashion accessories?

A: Handbags **B:** Scarves

C: Hats **D:** Shoes

59

Balmoral Castle was built for which monarch?

A: Anne **B:** Mary II

C: Elizabeth I **D:** Victoria

60

The industrial community of Saltaire was built near which city?

A: Birmingham **B:** Bradford

C: Bath **D:** Bristol

If you would like to use your 50:50 please turn to page 269
To ask the audience please turn to page 289
Turn to the answer section on page 304 to find out if you've won £8,000!

8 ◆ £8,000

61

The Palace Pier and West Pier are features of which English resort?

- **A:** Eastbourne
- **B:** Blackpool
- **C:** Skegness
- **D:** Brighton

62

To which royal house did Richard III belong?

- **A:** Tudor
- **B:** Stuart
- **C:** York
- **D:** Lancaster

63

What kind of creature is the anteater?

- **A:** Mammal
- **B:** Amphibian
- **C:** Reptile
- **D:** Fish

64

At which port was Brunel's ship Great Western built and launched?

- **A:** Bristol
- **B:** Portsmouth
- **C:** Glasgow
- **D:** Belfast

65

Which of these people was alive in the year 1900?

- **A:** Mother Teresa
- **B:** Yuri Gagarin
- **C:** Giacomo Puccini
- **D:** Charles Dickens

If you would like to use your 50:50 please turn to page 269
To ask the audience please turn to page 289
Turn to the answer section on page 304 to find out if you've won £8,000!

8 ◆ £8,000

Where is the most northerly
point on the British mainland?

A: Dunnet Head

B: Dungeness

C: Dunstable

D: Dundee

Which two numbers denote
normal vision in a human being?

A: 10/10

B: 20/20

C: 50/50

D: 80/80

What is the name of the small, red,
wild flower that closes in cloudy or rainy weather?

A: Scarlet pimpernel

B: Scarlet fever

C: Scarlet runner

D: Scarlet woman

What was the first name of Lady Hamilton,
who became a mistress of Lord Nelson?

A: Emily

B: Emilia

C: Emmeline

D: Emma

What is the reddish orange roe of a scallop called?

A: Shingle

B: Pearl

C: Laguna

D: Coral

If you would like to use your 50:50 please turn to page 269
To ask the audience please turn to page 289
Turn to the answer section on page 304 to find out if you've won £8,000!

8 ◆ £8,000

71

Which American film star is an alternative name for an inflatable life jacket?

A: Betty Grable
B: Mae West
C: Jane Russell
D: Marilyn Monroe

72

What type of bird is an albatross?

A: Woodland bird
B: Garden bird
C: Sea bird
D: Moorland bird

73

What type of food product is Monterey Jack?

A: Hard cheese
B: Sweet biscuit
C: Smoked sausage
D: Spiced beef

74

Which of these is an alternative name for the card game 'Beggar My Neighbour'?

A: Strip Jack Naked
B: In the Altogether
C: Birthday Suit Fours
D: Snip Snap Snorum

75

Which of these Hollywood film actresses was born in Switzerland?

A: Elke Sommer
B: Jane Fonda
C: Ursula Andress
D: Isabelle Adjani

If you would like to use your 50:50 please turn to page 269
To ask the audience please turn to page 289
Turn to the answer section on page 304 to find out if you've won £8,000!

9 ◆ £16,000

1

In Victorian England, what was a fingersmith?

A: Pianist

B: Postman

C: Pickpocket

D: Policeman

2

Who wrote the poem which begins
'Do not go gentle into that good night'?

A: Wilfred Owen

B: Ted Hughes

C: W.H. Auden

D: Dylan Thomas

3

In Germany, what kind of building is the 'Rathaus'?

A: Cathedral

B: Hospital

C: Town hall

D: Police Station

4

Which engineer designed the original
Paddington Station in London?

A: Thomas Telford

B: Isambard Kingdom Brunel

C: George Stephenson

D: Richard Trevithick

5

What is traditionally given as a thirtieth
wedding anniversary gift?

A: Crystal

B: Coral

C: Pearl

D: Ruby

If you would like to use your 50:50 please turn to page 269
To ask the audience please turn to page 289
Turn to the answer section on page 304 to find out if you've won £16,000!

9 ◆ £16,000

6

In which year did US military involvement
in the Vietnam War officially end?

A: 1970   **B:** 1973

C: 1975   **D:** 1979

7

On which late 1980s Saturday morning TV show
was Gordon the Gopher a regular fixture?

A: Swap Shop   **B:** Going Live!

C: It's Wicked!   **D:** Get Fresh

8

Which country does not participate in
rugby's annual Six Nations Championship?

A: France   **B:** Spain

C: Scotland   **D:** Italy

9

Who was the queen consort to King George V?

A: Mary   **B:** Alexandra

C: Adelaide   **D:** Charlotte

10

In which George Eliot novel does the
character Tertius Lydgate appear?

A: Daniel Deronda   **B:** Adam Bede

C: Middlemarch   **D:** Silas Marner

If you would like to use your 50:50 please turn to page 269
To ask the audience please turn to page 289
Turn to the answer section on page 304 to find out if you've won £16,000!

11

How many stars are there on the national flag of Australia?

A: 4

B: 6

C: 7

D: 9

12

In Greek mythology, Atlas was a member of which race?

A: Titans

B: Cyclops

C: Centaurs

D: Gorgons

13

Alicante is on which Spanish coast?

A: Costa Brava

B: Costa Blanca

C: Costa de la Luz

D: Costa del Sol

14

How many sisters made up the Pointer Sisters?

A: 2

B: 3

C: 4

D: 6

15

The wine Tokay originated in which country?

A: Germany

B: Hungary

C: Japan

D: Bulgaria

If you would like to use your 50:50 please turn to pages 269 and 270
To ask the audience please turn to page 290
Turn to the answer section on page 304 to find out if you've won £16,000!

9 ◆ £16,000

16

In which country did what is generally regarded as the first training school for guide dogs for the blind open in 1916?

A: Belgium

B: United Kingdom

C: United States

D: Germany

17

Where would a stevedore be most likely to work?

A: Dock

B: Railway

C: Hospital

D: Courtroom

18

The BBFC is concerned with what?

A: Film

B: Food

C: Football

D: Farming

19

Which of these buildings is featured as Hogwarts School in the 'Harry Potter' films?

A: Alnwick Castle

B: Chatsworth House

C: Hardwick Hall

D: Fountains Abbey

20

Battle Abbey was built to commemorate which battle?

A: Agincourt

B: Blenheim

C: Hastings

D: Trafalgar

If you would like to use your 50:50 please turn to page 270
To ask the audience please turn to page 290
Turn to the answer section on page 304 to find out if you've won £16,000!

9 ◆ £16,000

21

What is absent from an aseptic room?

A: Sound

B: Germs

C: Air

D: Light

22

Lillee and Thompson were fast bowlers for which country?

A: Australia

B: England

C: New Zealand

D: South Africa

23

'Golden lion' and 'emperor' are types of which primate?

A: Macaque

B: Tamarin

C: Marmoset

D: Baboon

24

What would you get if you ordered 'canard' in a French restaurant?

A: Chicken

B: Duck

C: Pork

D: Veal

25

Complete the title of the Hemingway novel: 'A Farewell to . . .'?

A: Love

B: England

C: Hope

D: Arms

If you would like to use your 50:50 please turn to page 270
To ask the audience please turn to page 290
Turn to the answer section on page 304 to find out if you've won £16,000!

26

Which of these gases is not an element?

- **A:** Chlorine
- **B:** Helium
- **C:** Nitrogen
- **D:** Carbon dioxide

27

Which battle brought the Tudor dynasty to power?

- **A:** Bosworth
- **B:** Hastings
- **C:** Flodden
- **D:** Naseby

28

Which of these actors was most famous for playing Dracula on film?

- **A:** William Powell
- **B:** Basil Rathbone
- **C:** Adam West
- **D:** Bela Lugosi

29

Which musical features the song 'Happy Talk'?

- **A:** Brigadoon
- **B:** Sunset Boulevard
- **C:** South Pacific
- **D:** Chicago

30

Stella Rimington was the first female head of which organisation?

- **A:** Football Association
- **B:** United Nations
- **C:** FBI
- **D:** MI5

If you would like to use your 50:50 please turn to page 270
To ask the audience please turn to page 290
Turn to the answer section on page 304 to find out if you've won £16,000!

31

What was the organisation formed by Charles de Gaulle in 1940?

A: New France

B: Free French

C: France Exile

D: London French

32

Which of these horses was a grey?

A: Nijinsky

B: Red Rum

C: Shergar

D: Desert Orchid

33

Which measurement is equivalent to 10,000 square metres?

A: Hectare

B: Acre

C: Rood

D: Centare

34

Which river flows through Lake Constance?

A: Rhine

B: Seine

C: Danube

D: Volga

35

The Jorvik Viking Centre is a tourist attraction in which city?

A: Bath

B: Exeter

C: Gloucester

D: York

If you would like to use your 50:50 please turn to page 270
To ask the audience please turn to page 290
Turn to the answer section on page 304 to find out if you've won £16,000!

9 ◆ £16,000

36

Who would be most likely to use
an orange stick in their work?

A: Dentist

B: Electrician

C: Manicurist

D: Teacher

37

Canada is divided into provinces and what else?

A: States

B: Counties

C: Shires

D: Territories

38

Which of these is made up of polyps?

A: Honeycomb

B: Cloud

C: Peat bog

D: Coral reef

39

Which of these animals is not an insectivore?

A: Mole

B: Hedgehog

C: Shrew

D: Weasel

40

Which entertainer is known as the Divine Miss M?

A: Madonna

B: Mariah Carey

C: Bette Midler

D: Kylie Minogue

If you would like to use your 50:50 please turn to page 270
To ask the audience please turn to page 290
Turn to the answer section on page 304 to find out if you've won £16,000!

9 ◆ £16,000

41

What is the usual habitat of the red grouse?

A: Woodland

B: Mountains

C: Seashore

D: Moorland

42

Which canal links the Yangtze and Yellow rivers?

A: Suez Canal

B: Kiel Canal

C: Grand Canal

D: Corinth Canal

43

Which of these US presidents was not a Republican?

A: Gerald Ford

B: Richard Nixon

C: George Bush Snr

D: Jimmy Carter

44

With which sport is Ian Balding most associated?

A: Cricket

B: Rugby union

C: Horse racing

D: Snooker

45

Which romantic novelist wrote the
Tilly Trotter series of novels?

A: Victoria Holt

B: Georgette Heyer

C: Barbara Cartland

D: Catherine Cookson

If you would like to use your 50:50 please turn to page 270
To ask the audience please turn to page 290
Turn to the answer section on page 304 to find out if you've won £16,000!

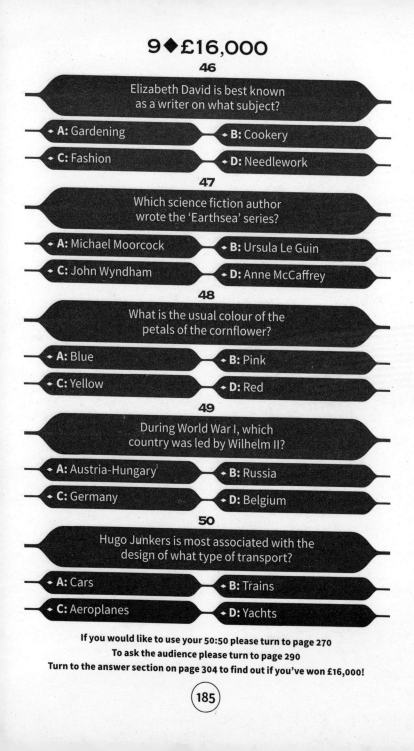

46

Elizabeth David is best known as a writer on what subject?

A: Gardening

B: Cookery

C: Fashion

D: Needlework

47

Which science fiction author wrote the 'Earthsea' series?

A: Michael Moorcock

B: Ursula Le Guin

C: John Wyndham

D: Anne McCaffrey

48

What is the usual colour of the petals of the cornflower?

A: Blue

B: Pink

C: Yellow

D: Red

49

During World War I, which country was led by Wilhelm II?

A: Austria-Hungary

B: Russia

C: Germany

D: Belgium

50

Hugo Junkers is most associated with the design of what type of transport?

A: Cars

B: Trains

C: Aeroplanes

D: Yachts

If you would like to use your 50:50 please turn to page 270

To ask the audience please turn to page 290

Turn to the answer section on page 304 to find out if you've won £16,000!

9 ◆ £16,000

51

The word 'safari' means journey in which language?

- **A:** Hindi
- **B:** Turkish
- **C:** Persian
- **D:** Swahili

52

In World War II, the Chindits fought in which country?

- **A:** Egypt
- **B:** Burma
- **C:** Japan
- **D:** France

53

Which of these rivers flows in a generally northwards direction?

- **A:** Volga
- **B:** Rhône
- **C:** Mississippi
- **D:** Nile

54

What kind of zodiac sign is Capricorn?

- **A:** Earth
- **B:** Air
- **C:** Fire
- **D:** Water

55

At which Winter Olympics did Torvill and Dean win gold?

- **A:** Lake Placid
- **B:** Sarajevo
- **C:** Lillehammer
- **D:** Albertville

If you would like to use your 50:50 please turn to page 270
To ask the audience please turn to page 290
Turn to the answer section on page 304 to find out if you've won £16,000!

9 ◆ £16,000

56

Which of these Middle Eastern countries is made up of islands?

A: Bahrain

B: Kuwait

C: Lebanon

D: Qatar

57

In the game of bar billiards, how many holes are set in the table's surface?

A: Three

B: Six

C: Nine

D: Twelve

58

Founded in Oxford, Blackwell's is best known for selling what?

A: Glassware

B: Books

C: Furniture

D: Theatrical costumes

59

Aldous Huxley's novel 'Brave New World' takes its title from which Shakespeare play?

A: Twelfth Night

B: Macbeth

C: Richard II

D: The Tempest

60

The British composer Eric Coates is associated with which famous march?

A: Colonel Bogey

B: The Dam Busters

C: Radetzky

D: Washington Post

If you would like to use your 50:50 please turn to page 270

To ask the audience please turn to page 290

Turn to the answer section on page 304 to find out if you've won £16,000!

61

Which of these islands celebrate
'Liberation Day' on May 9th?

- **A:** Channel Islands
- **B:** Isles of Scilly
- **C:** Farne Islands
- **D:** Orkney Islands

62

Which actress was nicknamed 'Hanoi Jane' after visiting
Vietnam and protesting against US involvement in that War?

- **A:** Jane Russell
- **B:** Jane Seymour
- **C:** Jane Fonda
- **D:** Jane Wyman

63

According to George Bernard Shaw, which two countries
'. . . are divided by a common language.'?

- **A:** Scotland & Ireland
- **B:** Spain & Mexico
- **C:** America & England
- **D:** France & Belgium

64

What was the name of Charlotte, Emily and
Anne Brontë's artistic brother, born in 1817?

- **A:** Baldwin
- **B:** Branwell
- **C:** Byron
- **D:** Bennet

65

Whose first album 'Tubular Bells' launched
Richard Branson's new record company in 1973?

- **A:** Bryan Ferry
- **B:** Rick Wakeman
- **C:** Steve Winwood
- **D:** Mike Oldfield

If you would like to use your 50:50 please turn to page 270
To ask the audience please turn to page 290
Turn to the answer section on page 304 to find out if you've won £16,000!

9♦£16,000

Abraham Darby's famous Ironbridge, near Coalbrookdale, spans which river?

A: Trent

B: Mersey

C: Avon

D: Severn

67

What was the first name of Berlioz, the nineteenth-century French composer?

A: Frédéric

B: Hector

C: Pierre

D: César

68

Which of these titles is a nickname for Chopin's famous Opus 64 No. 1?

A: Second Nocturne

B: Minute Waltz

C: Hour Concerto

D: Holy Day Symphony

69

Dotheboys Hall is a school that features in which Dickens novel?

A: Nicholas Nickleby

B: David Copperfield

C: Oliver Twist

D: Great Expectations

70

What was the profession of the American Aaron Copland?

A: Composer

B: Fashion designer

C: Baseball player

D: Film actor

If you would like to use your 50:50 please turn to pages 270 and 271
To ask the audience please turn to pages 290 and 291
Turn to the answer section on page 304 to find out if you've won £16,000!

9 ◆ £16,000

Which of these capital cities should be written
with the initials A.C.T. after its name?

A: Washington

B: Ottawa

C: Valetta

D: Canberra

Which English monarch was known as 'The Virgin Queen'?

A: Mary I

B: Elizabeth I

C: Anne

D: Victoria

Which of these words is the
collective term for a group of owls?

A: Parliament

B: Government

C: Cabinet

D: Ministry

Philip Mountbatten became
which of these people in 1947?

A: President of the USA

B: Duke of Edinburgh

C: Prince of Wales

D: King of Greece

The abbreviation V-TOL applies
to which type of transport?

A: Railway locomotive

B: Luxury coach

C: Cargo ship

D: Jump jet

If you would like to use your 50:50 please turn to page 271
To ask the audience please turn to page 291
Turn to the answer section on page 304 to find out if you've won £16,000!

1

For which country did Derek Quinnell and his sons Scott and Craig play rugby union?

A: England

B: Ireland

C: Scotland

D: Wales

2

In which county is Alderley Edge?

A: Cumbria

B: Northumberland

C: Durham

D: Cheshire

3

Of which medical complaint is 'variola' the Latin name?

A: Smallpox

B: Varicose vein

C: Migraine

D: Scarlet fever

4

In which city is the 1949 film 'The Third Man' set?

A: Paris

B: Vienna

C: Lisbon

D: Athens

5

Salicylic acid, used in aspirin, derives its name from the Latin for which tree?

A: Poplar

B: Sycamore

C: Tea tree

D: Willow

If you would like to use your 50:50 please turn to page 271
To ask the audience please turn to page 291
Turn to the answer section on page 304 to find out if you've won £32,000!

10 ◆ £32,000

6

In which sport did the Colombian
Juan Pablo Montoya find fame?

A: Tennis

B: Football

C: Motor racing

D: Golf

7

Which civilization lived in the city of Machu Picchu?

A: Maori

B: Inca

C: Aztec

D: Apache

8

Which of the following did Winston Churchill
describe as 'the end of the beginning'?

A: D-Day

B: Stalingrad

C: Battle of Britain

D: El Alamein

9

Which of these rivers is regarded
as the most sacred to Hindus?

A: Krishna

B: Ganges

C: Luni

D: Brahmaputra

10

Who wrote the song 'Jealous Guy', which in 1981
became the only UK No 1 for Roxy Music?

A: John Lennon

B: Mick Jagger

C: Ray Davies

D: Paul McCartney

If you would like to use your 50:50 please turn to page 271
To ask the audience please turn to page 291
Turn to the answer section on page 304 to find out if you've won £32,000!

10 ◆ £32,000

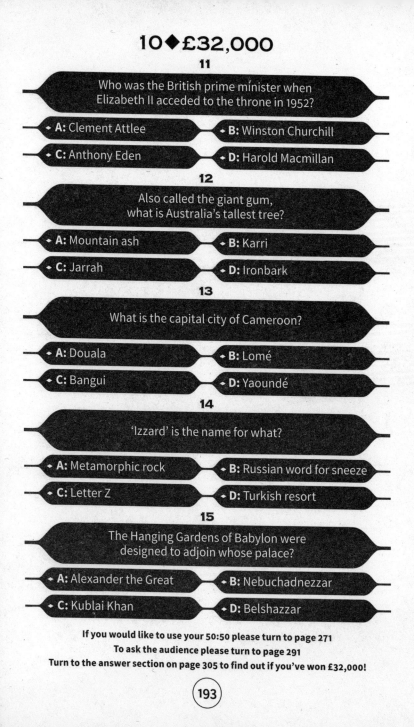

11

Who was the British prime minister when
Elizabeth II acceded to the throne in 1952?

A: Clement Attlee   **B:** Winston Churchill

C: Anthony Eden   **D:** Harold Macmillan

12

Also called the giant gum,
what is Australia's tallest tree?

A: Mountain ash   **B:** Karri

C: Jarrah   **D:** Ironbark

13

What is the capital city of Cameroon?

A: Douala   **B:** Lomé

C: Bangui   **D:** Yaoundé

14

'Izzard' is the name for what?

A: Metamorphic rock   **B:** Russian word for sneeze

C: Letter Z   **D:** Turkish resort

15

The Hanging Gardens of Babylon were
designed to adjoin whose palace?

A: Alexander the Great   **B:** Nebuchadnezzar

C: Kublai Khan   **D:** Belshazzar

If you would like to use your 50:50 please turn to page 271
To ask the audience please turn to page 291
Turn to the answer section on page 305 to find out if you've won £32,000!

10 ◆ £32,000

16

Which ancient mathematician was killed by a Roman soldier during the siege of Syracuse?

A: Pythagoras

B: Euclid

C: Archimedes

D: Zeno

17

What is the name of Bogota's main international airport?

A: El Dorado

B: Villareal

C: Santa Marta

D: Gorgona

18

In which sport is the Corbillon Cup a major competition?

A: Squash

B: Golf

C: Hockey

D: Table tennis

19

In which US state is the Augusta National golf course?

A: North Carolina

B: Florida

C: Georgia

D: Virginia

20

Who wrote the novel 'The Good Companions'?

A: E.M. Forster

B: J.B. Priestley

C: C.S. Lewis

D: D.H. Lawrence

If you would like to use your 50:50 please turn to page 271
To ask the audience please turn to page 291
Turn to the answer section on page 305 to find out if you've won £32,000!

21

The actress Marilyn Monroe
was married to which playwright?

A: Tennessee Williams | **B:** Eugene O'Neill

C: Arthur Miller | **D:** David Mamet

22

Which footballer wrote the
autobiography '1966 and All That'?

A: Gordon Banks | **B:** Jack Charlton

C: Geoff Hurst | **D:** Nobby Stiles

23

Who succeeded Churchill as prime minister in 1955?

A: Baldwin | **B:** Macmillan

C: Wilson | **D:** Eden

24

Yellowstone National Park lies chiefly in which US state?

A: North Dakota | **B:** Kansas

C: Wyoming | **D:** Nevada

25

Indonesia declared independence
from which country in 1945?

A: Portugal | **B:** France

C: Netherlands | **D:** Spain

If you would like to use your 50:50 please turn to page 271
To ask the audience please turn to page 291
Turn to the answer section on page 305 to find out if you've won £32,000!

10 ◆ £32,000

26

Which part of the face could best be described as a 'proboscis'?

- **A:** Chin
- **B:** Nose
- **C:** Eye
- **D:** Cheek

27

What was the name of Britain's first nuclear power station?

- **A:** Dounreay
- **B:** Calder Hall
- **C:** Chapelcross
- **D:** Sizewell

28

What is the first name of the fictional pilot known as 'Biggles'?

- **A:** William
- **B:** James
- **C:** Henry
- **D:** Edward

29

In which service does a petty officer work?

- **A:** Army
- **B:** Navy
- **C:** Air Force
- **D:** Police

30

What kind of person would be described as a 'prestidigitator'?

- **A:** Magician
- **B:** Fireman
- **C:** Teacher
- **D:** Artist

If you would like to use your 50:50 please turn to page 271

To ask the audience please turn to page 291

Turn to the answer section on page 305 to find out if you've won £32,000!

31

Which of these is a tool with a large curved blade, used for cutting grass or crops?

A: Blithe

B: Hythe

C: Writhe

D: Scythe

32

In which city is the Dome of the Rock?

A: Damascus

B: Alexandria

C: Baghdad

D: Jerusalem

33

Which of these colours appears on the flag of Greece?

A: Green

B: Blue

C: Yellow

D: Red

34

What name is given to the swampy Caribbean coast of Honduras and Nicaragua?

A: Coromandel Coast

B: Ivory Coast

C: Gold Coast

D: Mosquito Coast

35

Carrots are a particularly good source of which vitamin?

A: Vitamin A

B: Vitamin D

C: Vitamin E

D: Vitamin K

If you would like to use your 50:50 please turn to page 271
To ask the audience please turn to page 291
Turn to the answer section on page 305 to find out if you've won £32,000!

10 ♦ £32,000

36

Where on its course is the
River Thames known as the Isis?

A: Eton

B: Henley

C: Putney

D: Oxford

37

Which of these was a forerunner of reggae?

A: Ska

B: Skat

C: Rap

D: Jungle

38

Barcelona is the capital of which Spanish region?

A: Andalusia

B: Catalonia

C: Galicia

D: La Rioja

39

Which TV presenter wrote the novel 'The Soldier's Return'?

A: Muriel Gray

B: Francine Stock

C: Melvyn Bragg

D: Alan Titchmarsh

40

If you had Montessori training, what would you be?

A: Priest

B: Teacher

C: Midwife

D: Pilot

If you would like to use your 50:50 please turn to page 271
To ask the audience please turn to page 291
Turn to the answer section on page 305 to find out if you've won £32,000!

10◆£32,000

41

Which actress is the daughter of Goldie Hawn?

A: Salma Hayek

B: Chloe Sevigny

C: Reese Witherspoon

D: Kate Hudson

42

Who wrote the play 'The Seagull'?

A: O'Neill

B: Strindberg

C: Chekhov

D: Shaw

43

Michael Barratt is the real name of which pop star?

A: Billy Idol

B: Suggs

C: Shakin' Stevens

D: Meatloaf

44

Gadshill, near Rochester, was the home of which novelist?

A: Charles Dickens

B: H.G. Wells

C: Virginia Woolf

D: D.H. Lawrence

45

With which war is the poet Rupert Brooke most associated?

A: Boer War

B: World War I

C: World War II

D: Korean War

If you would like to use your 50:50 please turn to page 271
To ask the audience please turn to page 291
Turn to the answer section on page 305 to find out if you've won £32,000!

10◆£32,000

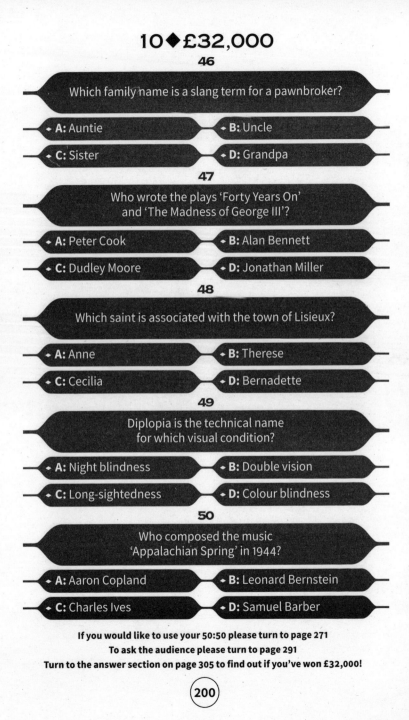

46

Which family name is a slang term for a pawnbroker?

- **A:** Auntie
- **B:** Uncle
- **C:** Sister
- **D:** Grandpa

47

Who wrote the plays 'Forty Years On' and 'The Madness of George III'?

- **A:** Peter Cook
- **B:** Alan Bennett
- **C:** Dudley Moore
- **D:** Jonathan Miller

48

Which saint is associated with the town of Lisieux?

- **A:** Anne
- **B:** Therese
- **C:** Cecilia
- **D:** Bernadette

49

Diplopia is the technical name for which visual condition?

- **A:** Night blindness
- **B:** Double vision
- **C:** Long-sightedness
- **D:** Colour blindness

50

Who composed the music 'Appalachian Spring' in 1944?

- **A:** Aaron Copland
- **B:** Leonard Bernstein
- **C:** Charles Ives
- **D:** Samuel Barber

If you would like to use your 50:50 please turn to page 271
To ask the audience please turn to page 291
Turn to the answer section on page 305 to find out if you've won £32,000!

10 ◆ £32,000

51

The Coral Sea is a part of which ocean?

A: Arctic

B: Pacific

C: Indian

D: Atlantic

52

What is the name of the world's highest active volcano?

A: Etna

B: St Helens

C: Cotopaxi

D: Krakatoa

53

By what name is the singer Florian Cloud De Bounevialle Armstrong known?

A: Dido

B: Gabrielle

C: Enya

D: Sade

54

Of which organisation was the Swede Dag Hammarskjold secretary-general?

A: Commonwealth

B: United Nations

C: Red Cross

D: NATO

55

In Greek mythology, which of these was a giant with a hundred eyes?

A: Argus

B: Daphnis

C: Agamemnon

D: Narcissus

If you would like to use your 50:50 please turn to page 272
To ask the audience please turn to page 292
Turn to the answer section on page 305 to find out if you've won £32,000!

10 ◆ £32,000

In which part of the world was the
Mayan civilisation situated?

A: Southern Africa

B: Eastern Asia

C: Central America

D: Northern Europe

Which European capital stands on the confluence
of the Danube and the Sava rivers?

A: Prague

B: Zagreb

C: Sofia

D: Belgrade

Which of these films did not star a married couple?

A: Shanghai Surprise

B: Eyes Wide Shut

C: Key Largo

D: Bringing Up Baby

Queen Mary II belonged to which British royal house?

A: York

B: Lancaster

C: Tudor

D: Stuart

Which rock star named himself after a hearing-aid shop?

A: Sting

B: Lemmy

C: Fish

D: Bono

If you would like to use your 50:50 please turn to page 272
To ask the audience please turn to page 292
Turn to the answer section on page 305 to find out if you've won £32,000!

10 ◆ £32,000

61

Most of the collection of the British Library
in London was formerly housed where?

A: British Museum

B: National Gallery

C: Fitzwilliam Museum

D: Buckingham Palace

62

What is the meaning of 'pianoforte',
the full name for the instrument we just call a piano?

A: High-low

B: Sharp-flat

C: Soft-loud

D: Left-right

63

Sirius is another name for what?

A: Dog Violet

B: Dog Star

C: Dog Fish

D: Dog Latin

64

Which of the states of the USA
is nicknamed 'The Empire State'?

A: New York

B: Rhode Island

C: New Jersey

D: Virginia

65

What is the meaning of the word 'austral'?

A: Southern

B: Opposite

C: Upside-down

D: Round the back

If you would like to use your 50:50 please turn to page 272
To ask the audience please turn to page 292
Turn to the answer section on page 305 to find out if you've won £32,000!

10 ◆ £32,000

66

The Smithsonian is an institution of
national museums in which city?

A: Toronto

B: Washington DC

C: Glasgow

D: London

67

Which silent film star was known as
'The Great Stone Face' because of his deadpan expression?

A: Harry Langdon

B: Buster Keaton

C: Fatty Arbuckle

D: Stan Laurel

68

Which of these Italian cities lies south of Rome?

A: Naples

B: Venice

C: Bologna

D: Turin

69

Which of these people, after whom articles
of clothing are named, is made up?

A: Jules Léotard

B: Levi Strauss

C: Carlos Tuxedo

D: Amelia Bloomer

70

Who was the composer of the song
'There's No Business Like Show Business'?

A: Jerome Kern

B: Irving Berlin

C: Cole Porter

D: George Gershwin

If you would like to use your 50:50 please turn to page 272

To ask the audience please turn to page 292

Turn to the answer section on page 305 to find out if you've won £32,000!

10 ◆ £32,000

71

Which of these entertainers was NOT born in England?

A: Charlie Chaplin

B: Bob Hope

C: Stand Laurel

D: Oliver Hardy

72

Which Biblical king was particularly known for his great wisdom?

A: Solomon

B: Saul

C: Stephen

D: Simon

73

Which of these refers to goods or wreckage lying on the seabed?

A: Flotsam

B: Jetsam

C: Lagan

D: Caragan

74

Who did Jack Ruby notoriously shoot on November 24th 1963?

A: John Fitzgerald Kennedy

B: Martin Luther King

C: Lee Harvey Oswald

D: James Earl Ray

75

Which statesman popularised the term 'The Iron Curtain', in a 1946 speech?

A: Dwight D. Eisenhower

B: Winston Churchill

C: Clement Atlee

D: Franklin D. Roosevelt

If you would like to use your 50:50 please turn to page 272
To ask the audience please turn to page 292
Turn to the answer section on page 305 to find out if you've won £32,000!

11 ◆ £64,000

1

What should a traditional Dundee cake not contain?

A: Sultanas

B: Mixed peel

C: Glacé cherries

D: Almonds

2

Who patented a motion picture camera in 1891?

A: Eastman

B: Edison

C: Dickson

D: Lumière

3

In Gilbert and Sullivan's 'The Mikado', who is the son of the Mikado?

A: Nanki-Poo

B: Poon-Ban

C: Yum-Yum

D: Pish-Tush

4

Which musical features the characters Rum Tum Tugger and Skimbleshanks?

A: Barnum

B: Cabaret

C: The Lion King

D: Cats

5

Who wrote the novels featuring British soldier Sharpe?

A: C.S. Forester

B: Patrick O'Brian

C: Bernard Cornwell

D: Joseph Conrad

If you would like to use your 50:50 please turn to page 272
To ask the audience please turn to page 292
Turn to the answer section on page 305 to find out if you've won £64,000!

6

Where was orange marmalade devised in the 18th century?

- **A:** Seville
- **B:** Dundee
- **C:** Oxford
- **D:** Périgord

7

Kublai Khan was emperor of which country?

- **A:** Egypt
- **B:** Persia
- **C:** India
- **D:** China

8

Who is the hero of James Fenimore Cooper's 'The Last of the Mohicans'?

- **A:** Matty Grumppo
- **B:** Natty Bumppo
- **C:** Ratty Humppo
- **D:** Fatty Lumppo

9

In the 'Wayne's World' films, who is Wayne's best friend?

- **A:** Bill S Preston
- **B:** Garth Algar
- **C:** Ted 'Theodore' Logan
- **D:** Oliver Woods

10

Teddy and Ruth are characters in which Harold Pinter play?

- **A:** The Birthday Party
- **B:** The Caretaker
- **C:** The Dumb Waiter
- **D:** The Homecoming

If you would like to use your 50:50 please turn to page 272

To ask the audience please turn to page 292

Turn to the answer section on page 305 to find out if you've won £64,000!

11 ◆ £64,000

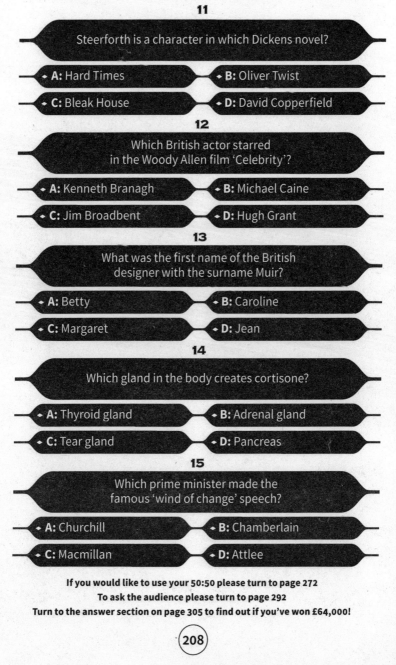

11

Steerforth is a character in which Dickens novel?

A: Hard Times

B: Oliver Twist

C: Bleak House

D: David Copperfield

12

Which British actor starred
in the Woody Allen film 'Celebrity'?

A: Kenneth Branagh

B: Michael Caine

C: Jim Broadbent

D: Hugh Grant

13

What was the first name of the British
designer with the surname Muir?

A: Betty

B: Caroline

C: Margaret

D: Jean

14

Which gland in the body creates cortisone?

A: Thyroid gland

B: Adrenal gland

C: Tear gland

D: Pancreas

15

Which prime minister made the
famous 'wind of change' speech?

A: Churchill

B: Chamberlain

C: Macmillan

D: Attlee

If you would like to use your 50:50 please turn to page 272
To ask the audience please turn to page 292
Turn to the answer section on page 305 to find out if you've won £64,000!

16

Which famous building is sometimes called the 'Flavian Amphitheatre'?

A: Acropolis

B: Colosseum

C: Taj Mahal

D: Topkapi Palace

17

What kind of animal is a 'pangolin'?

A: Mammal

B: Reptile

C: Amphibian

D: Bird

18

The Vedas are sacred texts of which religion?

A: Hinduism

B: Islam

C: Buddhism

D: Judaism

19

Who composed the 'Peer Gynt' suite?

A: Sibelius

B: Handel

C: Dvořák

D: Grieg

20

With which band is Ozzy Osbourne most associated?

A: Def Leppard

B: Black Sabbath

C: Motorhead

D: Judas Priest

If you would like to use your 50:50 please turn to page 272

To ask the audience please turn to page 292

Turn to the answer section on page 305 to find out if you've won £64,000!

21

Thomas Becket was murdered during the reign of which king?

A: Henry I

B: Henry II

C: Edward I

D: Edward II

22

What kind of material is 'morocco'?

A: Leather

B: Felt

C: Linen

D: Cotton

23

Who famously lived at Buckland Abbey in Devon?

A: Charles Darwin

B: Francis Drake

C: Rudyard Kipling

D: Cardinal Wolsey

24

The Marshall Plan was intended to help countries on which continent after World War II?

A: Africa

B: Europe

C: Asia

D: South America

25

What is the official language of Pakistan?

A: Farsi

B: Urdu

C: Bengali

D: Gujarati

If you would like to use your 50:50 please turn to page 272

To ask the audience please turn to page 292

Turn to the answer section on page 305 to find out if you've won £64,000!

11 ◆ £64,000

In which year was Abraham Lincoln assassinated?

A: 1865

B: 1875

C: 1885

D: 1895

27

What kind of flower is a lady's slipper?

A: Orchid

B: Rose

C: Lily

D: Petunia

28

What kind of clothing is a 'beanie'?

A: Boot

B: Shirt

C: Hat

D: Glove

29

Bobby Charlton was briefly the manager of which northern football club?

A: Bolton Wanderers

B: Tranmere Rovers

C: Crewe Alexandra

D: Preston North End

30

Pomerania is a region on which continent?

A: Asia

B: Europe

C: Africa

D: South America

If you would like to use your 50:50 please turn to page 272
To ask the audience please turn to page 292
Turn to the answer section on page 305 to find out if you've won £64,000!

11 ◆ £64,000

31

Ewart was the middle name of
which prime minister?

A: Disraeli   **B:** Gladstone

C: Canning   **D:** Palmerston

32

Which of these countries does not have
land on the island of Borneo?

A: Indonesia   **B:** Brunei

C: Malaysia   **D:** China

33

At Cambridge, what is the name of the lawns
which slope down to the river?

A: The Rounds   **B:** The Backs

C: The Lefts   **D:** The Rights

34

Which part of the body is affected by 'quinsy'?

A: Throat   **B:** Stomach

C: Liver   **D:** Heart

35

The Cook Islands form a territory of which country?

A: Papua New Guinea   **B:** Great Britain

C: Australia   **D:** New Zealand

If you would like to use your 50:50 please turn to pages 272 and 273
To ask the audience please turn to page 293
Turn to the answer section on page 305 to find out if you've won £64,000!

36

Sutton Hoo in Suffolk is most associated with which archaeological discovery of 1939?

A: Stone age pottery

B: Viking helmet

C: Norman abbey

D: Saxon ship

37

Sir John Vanbrugh was famous as an architect and what else?

A: Doctor

B: Playwright

C: Bishop

D: Politician

38

What was the first name of the Spanish artist Goya?

A: Javier

B: Miguel

C: Francisco

D: Pedro

39

Who was the first English monarch to use the name Great Britain in her title, after England united with Scotland?

A: Mary II

B: Anne

C: Victoria

D: Elizabeth II

40

Which subject '. . . is more or less bunk' according to the famous quote by Henry Ford?

A: Chemistry

B: History

C: Psychology

D: Geography

If you would like to use your 50:50 please turn to page 273
To ask the audience please turn to page 293
Turn to the answer section on page 305 to find out if you've won £64,000!

11 ◆ £64,000

41

Which of these words refers to the northern parts of the world?

- **A:** Diurnal
- **B:** Hibernal
- **C:** Boreal
- **D:** Aestival

42

In which form did Nostradamus write his predictions?

- **A:** Blank verse
- **B:** Rhyming couplets
- **C:** Acrostics
- **D:** Quatrains

43

What is a 'clerihew'?

- **A:** Short comic verse
- **B:** Irish pixie
- **C:** Small drum
- **D:** Curved walking stick

44

Which of these films is based on a novel by C.S. Forester?

- **A:** The Italian Job
- **B:** The Odessa File
- **C:** The African Queen
- **D:** The Maltese Falcon

45

Who founded the first regular Atlantic steamship line?

- **A:** Thomas Cook
- **B:** Samuel Cunard
- **C:** Henry Bell
- **D:** Ian Robey

If you would like to use your 50:50 please turn to page 273
To ask the audience please turn to page 293
Turn to the answer section on page 305 to find out if you've won £64,000!

46

Shakespeare's play 'The Taming of the Shrew' was the inspiration for which musical?

◆ **A:** Funny Girl

B: Kiss Me Kate

◆ **C:** The Pajama Game

D: Kismet

47

The 'Lords spiritual' are the bishops in the House of Lords, so how are the others referred to?

◆ **A:** Lords temporal

B: Lords sepulchral

◆ **C:** Lords nominal

D: Lords dimensional

48

Who wrote the music to the song 'Mack the Knife'?

◆ **A:** Cole Porter

B: George Gershwin

◆ **C:** Richard Rodgers

D: Kurt Weill

49

According to legend, what did Joseph of Arimathea bring to Glastonbury?

◆ **A:** Noah's Ark

B: Crown of Thorns

◆ **C:** Turin Shroud

D: Holy Grail

50

What was described by Dr Johnson as worth seeing, but not worth going to see?

◆ **A:** Giant's Causeway

B: Taj Mahal

◆ **C:** Great Pyramid

D: Grand Canyon

If you would like to use your 50:50 please turn to page 273
To ask the audience please turn to page 293
Turn to the answer section on page 305 to find out if you've won £64,000!

12 ◆ £125,000

1

Who composed the score for the film 'The Piano'?

- **A:** John Williams
- **B:** Michael Nyman
- **C:** John Barry
- **D:** Elmer Bernstein

2

In cycling parlance, what are 'snakebites'?

- **A:** Punctures
- **B:** Chain cuts
- **C:** Grazes
- **D:** Collisions

3

In which city was Zsa Zsa Gabor born?

- **A:** Vienna
- **B:** Prague
- **C:** Bratislava
- **D:** Budapest

4

'An Awfully Big Adventure' is a novel by which author?

- **A:** Kate Atkinson
- **B:** P.D. James
- **C:** Beryl Bainbridge
- **D:** Daphne du Maurier

5

Which Roman emperor was the grandson of Livia and the nephew of Tiberius?

- **A:** Claudius
- **B:** Caligula
- **C:** Nerva
- **D:** Gallus

If you would like to use your 50:50 please turn to page 273
To ask the audience please turn to page 293
Turn to the answer section on page 305 to find out if you've won £125,000!

12 ◆ £125,000

6

Who was the first husband of actress Joan Collins?

- **A:** Anthony Newley
- **B:** Lionel Bart
- **C:** Oliver Reed
- **D:** Maxwell Reed

7

Who organised a 1568 plot to kill Elizabeth I?

- **A:** Anthony Babington
- **B:** Francis Drake
- **C:** John Cole
- **D:** Peter Bales

8

An oenophile is a connoisseur of what?

- **A:** Wines
- **B:** Cigars
- **C:** Paintings
- **D:** Operas

9

In which year were the first euro banknotes introduced into the European Union?

- **A:** 2000
- **B:** 2001
- **C:** 2002
- **D:** 2003

10

The 1979 film 'The Rose' was loosely based on the life and death of which singer?

- **A:** Billie Holliday
- **B:** Janis Joplin
- **C:** Judy Garland
- **D:** Alma Cogan

If you would like to use your 50:50 please turn to page 273
To ask the audience please turn to page 293
Turn to the answer section on page 305 to find out if you've won £125,000!

11

What kind of creature is a bulbul?

A: Bird

B: Antelope

C: Monkey

D: Snake

12

Which Jane Austen character owns Donwell Abbey and estate?

A: Mr Darcy

B: Captain Wentworth

C: Sir Thomas Bertram

D: Mr Knightley

13

What is the literal translation of the cocktail pina colada?

A: Crushed pineapple

B: Cold pineapple

C: Strained pineapple

D: Iced pineapple

14

Which explorer was the first to lead an expedition across the Pacific Ocean?

A: John Cabot

B: Christopher Columbus

C: Ferdinand Magellan

D: Vasco da Gama

15

Which acid forms the basis of aspirin?

A: Acetic acid

B: Amino acid

C: Salicylic acid

D: Lactic acid

If you would like to use your 50:50 please turn to page 273
To ask the audience please turn to page 293
Turn to the answer section on page 306 to find out if you've won £125,000!

12 ◆ £125,000

16

Which of these animals would be classed as a cetacean?

- **A:** Whale
- **B:** Seal
- **C:** Alligator
- **D:** Porcupine

17

What kind of bird is a rockhopper?

- **A:** Gull
- **B:** Penguin
- **C:** Owl
- **D:** Pigeon

18

Which of these is the home of the Dukes of Beaufort?

- **A:** Badminton
- **B:** Chatsworth
- **C:** Woburn
- **D:** Althorp

19

Yucatan is a state in which country?

- **A:** Panama
- **B:** Guatemala
- **C:** Mexico
- **D:** Honduras

20

In which century was Sir Isaac Newton born?

- **A:** 14th
- **B:** 15th
- **C:** 16th
- **D:** 17th

If you would like to use your 50:50 please turn to page 273
To ask the audience please turn to page 293
Turn to the answer section on page 306 to find out if you've won £125,000!

12 ◆ £125,000

21

Le Corbusier is a famous name in which field?

A: Architecture

B: Finance

C: Opera

D: Industry

22

The concept of Nirvana is associated with which of these religions?

A: Buddhism

B: Sikhism

C: Christianity

D: Islam

23

What kind of geographical feature is a 'strath'?

A: Valley

B: Island

C: River

D: Lake

24

In weaving, which thread runs in the opposite direction to the weft?

A: Wing

B: Welt

C: Warp

D: Wurl

25

Which king of England was murdered in Berkeley Castle?

A: Richard II

B: William II

C: Edward II

D: Henry II

If you would like to use your 50:50 please turn to page 273
To ask the audience please turn to page 293
Turn to the answer section on page 306 to find out if you've won £125,000!

12 ◆ £125,000

26

Britain's Charlie Magri was a
world champion in which sport?

A: Snooker
B: Darts
C: Boxing
D: Archery

27

Who wrote the novel 'The English Patient'?

A: Louis de Bernieres
B: Sebastian Faulkes
C: Michael Ondaatje
D: Kazuo Ishiguro

28

What kind of bird is a gadwall?

A: Duck
B: Heron
C: Pigeon
D: Vulture

29

Kodiak Island is part of which country?

A: Canada
B: Japan
C: United States
D: Russia

30

Which British king married Mary of Teck?

A: William I
B: Henry VIII
C: Charles II
D: George V

If you would like to use your 50:50 please turn to page 273
To ask the audience please turn to page 293
Turn to the answer section on page 306 to find out if you've won £125,000!

12 ◆ £125,000

31

Which of these cities shares its name
with the river on which it stands?

A: Adelaide

B: Brisbane

C: Hobart

D: Perth

32

What name is often used to describe
a male red deer over five years old?

A: Hare

B: Hinny

C: Hinky

D: Hart

33

Which stone provided the
key to understanding hieroglyphics?

A: Zapetta Stone

B: Andretta Stone

C: Rosetta Stone

D: Canetta Stone

34

The poet Sir Philip Sidney lived mainly
during the reign of which monarch?

A: Henry V

B: George III

C: Elizabeth I

D: Victoria

35

What kind of art work is 'Bargello'?

A: Embroidery

B: Porcelain

C: Glass

D: Watercolours

If you would like to use your 50:50 please turn to page 273
To ask the audience please turn to page 293
Turn to the answer section on page 306 to find out if you've won £125,000!

12 ◆ £125,000

36

What is the name of Australia's most important horse race?

A: Canberra Trophy

B: Sydney Guineas

C: Melbourne Cup

D: Adelaide Stakes

37

The Amalienborg Palace is in which capital city?

A: Helsinki

B: Stockholm

C: Oslo

D: Copenhagen

38

Which leader was found murdered, shortly after his wedding, in 453 AD?

A: Genghis Khan

B: William the Conqueror

C: Hannibal

D: Attila the Hun

39

Which London theatre was the first to be lit by electricity?

A: Shaftesbury

B: Theatre Royal

C: Savoy

D: Dominion

40

Which of these birds could be described as 'corvine'?

A: Peacock

B: Dove

C: Raven

D: Kingfisher

If you would like to use your 50:50 please turn to pages 273 and 274
To ask the audience please turn to pages 293 and 294
Turn to the answer section on page 306 to find out if you've won £125,000!

12 ◆ £125,000

41

Which of these words is derived from the name of legendary Scandinavian warriors?

A: Amok   **B:** Berserk

C: Chaos   **D:** Delirium

42

Which of these conditions is caused by a deficiency of vitamin B1?

A: Rickets   **B:** Scurvy

C: Beriberi   **D:** Malaria

43

In 1541, Hernando de Soto led the first group of Europeans to see which river?

A: Ganges   **B:** Niger

C: Mississippi   **D:** Amazon

44

The auditorium ceiling of the Paris Opera House was painted by which artist?

A: Renoir   **B:** Chagall

C: Picasso   **D:** Manet

45

The medieval monk the Venerable Bede lived most of his adult life at which monastery?

A: Rievaulx   **B:** Jarrow

C: Iona   **D:** Fountains Abbey

If you would like to use your 50:50 please turn to page 274
To ask the audience please turn to page 294
Turn to the answer section on page 306 to find out if you've won £125,000!

12 ◆ £125,000

46

Which report formed the basis
of the welfare state in Britain?

- **A:** Beveridge
- **B:** Scott
- **C:** Beeching
- **D:** Taylor

47

Which of the Channel Islands has an
hereditary overlord known as the 'Seigneur'?

- **A:** Jersey
- **B:** Guernsey
- **C:** Alderney
- **D:** Sark

48

Which people rule in a 'stratocracy'?

- **A:** Religious clerics
- **B:** The Mob
- **C:** Scientists
- **D:** The Military

49

Which art gallery is part of
London's South Bank complex?

- **A:** Wallace Collection
- **B:** National Gallery
- **C:** Hayward Gallery
- **D:** Serpentine Gallery

50

William Holman Hunt is most associated
with which group of artists?

- **A:** Cubists
- **B:** Surrealists
- **C:** Pop artists
- **D:** Pre-Raphaelites

If you would like to use your 50:50 please turn to page 274
To ask the audience please turn to page 294
Turn to the answer section on page 306 to find out if you've won £125,000!

13 ◆ £250,000

1

Which scientist is credited with discovering that blood circulates in the body?

A: William Gilbert

B: Robert Boyle

C: Isaac Newton

D: William Harvey

2

What is the literal meaning of 'Luftwaffe'?

A: Air strike

B: Air weapon

C: Air strength

D: Air force

3

What is the largest lake in Africa?

A: Tanganyika

B: Victoria

C: Chad

D: Nasser

4

What was the intended name of Beethoven's only opera 'Fidelio'?

A: Leonore

B: Alicia

C: Clara

D: Eleanor

5

The first Modern Olympic Games were held in Athens – in which year?

A: 1892

B: 1896

C: 1900

D: 1904

If you would like to use your 50:50 please turn to page 274
To ask the audience please turn to page 294
Turn to the answer section on page 306 to find out if you've won £250,000!

13 ◆ £250,000

6

Which London theatre was built to stage the works of Gilbert and Sullivan?

- **A:** Haymarket
- **B:** Savoy
- **C:** Palladium
- **D:** Lyceum

7

Which Russian ballet star defected to the West in September 1970?

- **A:** Baryshnikov
- **B:** Nureyev
- **C:** Pavlova
- **D:** Makarova

8

Who wrote the lyrics for the song 'I Could Write a Book'?

- **A:** Oscar Hammerstein
- **B:** Irving Berlin
- **C:** Lorenz Hart
- **D:** Cole Porter

9

For which feat of engineering is Joseph Bazalgette best known?

- **A:** London sewers
- **B:** Humber Bridge
- **C:** Flying Scotsman
- **D:** Concorde

10

What nationality was the scientist Anders Celsius?

- **A:** German
- **B:** Dutch
- **C:** Swedish
- **D:** Polish

If you would like to use your 50:50 please turn to page 274
To ask the audience please turn to page 294
Turn to the answer section on page 306 to find out if you've won £250,000!

13 ◆ £250,000

11

Who wrote the poem 'Kubla Khan'?

A: Byron
B: Coleridge
C: Shelley
D: Wordsworth

12

Which of these buildings was founded by Edward the Confessor?

A: Fountains Abbey
B: St Paul's Cathedral
C: Westminster Abbey
D: York Minster

13

Who was the first European navigator to reach India by sea?

A: Christopher Columbus
B: Vasco da Gama
C: Francis Drake
D: John Cabot

14

Which British composer wrote 'The Lark Ascending'?

A: Britten
B: Holst
C: Delius
D: Vaughan Williams

15

What nationality was the medieval saint Thomas Aquinas?

A: Italian
B: Spanish
C: French
D: Irish

If you would like to use your 50:50 please turn to page 274
To ask the audience please turn to page 294
Turn to the answer section on page 306 to find out if you've won £250,000!

13 ◆ £250,000

What word means hibernation during the summer?

A: Hermetism

B: Aestivation

C: Somnambulism

D: Thermotropism

17

Who is the patron saint of mountaineers?

A: Gotthard

B: Bernard

C: Sebastian

D: Moritz

18

Which of these artists is famous
for paintings of Californian swimming pools?

A: Andy Warhol

B: David Hockney

C: Roy Lichtenstein

D: Francis Bacon

19

Vladimir Horowitz was a virtuoso on which instrument?

A: Violin

B: Piano

C: Cello

D: Oboe

20

A haiku is a poem consisting of how many syllables?

A: 17

B: 21

C: 37

D: 41

If you would like to use your 50:50 please turn to page 274
To ask the audience please turn to page 294
Turn to the answer section on page 306 to find out if you've won £250,000!

13 ◆ £250,000

21

Which Commonwealth city was named
after the ancient name for Edinburgh?

- **A:** Canberra
- **B:** Toronto
- **C:** Dunedin
- **D:** Pretoria

22

Dame Myra Hess was a virtuoso on
which musical instrument?

- **A:** Cello
- **B:** Harp
- **C:** Violin
- **D:** Piano

23

Cape Breton Island is part of which Canadian province?

- **A:** Prince Edward Island
- **B:** Nova Scotia
- **C:** Quebec
- **D:** New Brunswick

24

Which of these countries does not
share a border with Guyana?

- **A:** Brazil
- **B:** Colombia
- **C:** Surinam
- **D:** Venezuela

25

What was the first name of the
furniture maker Hepplewhite?

- **A:** Josiah
- **B:** William
- **C:** George
- **D:** Thomas

If you would like to use your 50:50 please turn to page 274
To ask the audience please turn to page 294
Turn to the answer section on page 306 to find out if you've won £250,000!

13◆£250,000

Which of these bishops is not automatically
a member of the House of Lords?

A: Winchester

B: Durham

C: London

D: Salisbury

27

What is the female equivalent
of the Oedipus complex?

A: Electra complex

B: Athena complex

C: Diana complex

D: Pandora complex

28

Which country is known by its native inhabitants
as 'Aotearoa', 'Land of the long white cloud'?

A: Australia

B: New Zealand

C: Chile

D: Nepal

29

From which type of wood did Noah build the Ark?

A: Lime

B: Gopher

C: Date palm

D: Cork oak

30

Crispin is the patron saint of which craftsmen?

A: Thatchers

B: Shoemakers

C: Coopers

D: Clockmakers

If you would like to use your 50:50 please turn to page 274
To ask the audience please turn to page 294
Turn to the answer section on page 306 to find out if you've won £250,000!

13◆£250,000

31

In which English city was the
aviator Amy Johnson born?

A: Rochdale   **B:** Hull

C: Lincoln   **D:** Nottingham

32

In Greek mythology, what did the Hesperides guard?

A: Gateway to the Underworld   **B:** Fountain of wisdom

C: Tree with golden apples   **D:** Home of the Gods

33

Which of these countries is one of the
world's largest producers of vanilla?

A: Mongolia   **B:** Argentina

C: Madagascar   **D:** Nigeria

34

The archaeologist Sir Arthur Evans was most
associated with the excavation of which ancient site?

A: Thebes   **B:** Knossos

C: Troy   **D:** Babylon

35

Which foreign monarch worked incognito
in the royal shipyards at Deptford?

A: Napoleon Bonaparte   **B:** Peter the Great

C: Louis XIV   **D:** Nicholas II

If you would like to use your 50:50 please turn to page 274
To ask the audience please turn to page 294
Turn to the answer section on page 306 to find out if you've won £250,000!

36

Who was the first chancellor of the exchequer in the Labour government of 1945?

A: Hugh Dalton

B: Stafford Cripps

C: Aneurin Bevan

D: Clement Attlee

37

What was the name of Jane Austen's sister, who was her closest companion?

A: Alexandra

B: Cassandra

C: Maria

D: Anna

38

In Sumo wrestling, what is the dohyo?

A: Referee

B: Loincloth

C: Clay ring

D: Pre-fight ceremony

39

In Norse mythology, which plant killed the god Baldur?

A: Holly

B: Ivy

C: Mistletoe

D: Rose

40

If vegetables are described as 'macédoine', how are they prepared?

A: Diced

B: Shredded

C: Peeled

D: Mashed

If you would like to use your 50:50 please turn to page 274
To ask the audience please turn to page 294
Turn to the answer section on page 306 to find out if you've won £250,000!

13◆£250,000

41

Which playwright was killed in a Deptford tavern in 1593?

A: John Webster

B: Ben Jonson

C: George Farquhar

D: Christopher Marlowe

42

In which of these films does Elvis presley play a boxer?

A: Kid Galahad

B: King Creole

C: Roustabout

D: Flaming Star

43

How many episodes of the cult detective series 'Police Squad!' were made?

A: Six

B: Four

C: Twelve

D: One

44

Which quiz show host directed the 1970s comedy 'Bless This House'?

A: Magnus Magnusson

B: Richard Whiteley

C: William G. Stewart

D: Bob Monkhouse

45

'The Queen of Spain's Beard' was the title of an episode from which classic sitcom?

A: Blackadder

B: Dad's Army

C: Fawlty Towers

D: The Young Ones

If you would like to use your 50:50 please turn to pages 274 and 275
To ask the audience please turn to pages 294 and 295
Turn to the answer section on page 306 to find out if you've won £250,000!

13 ◆ £250,000

46

Which monarch was killed at the battle of Flodden field?

A: James IV of Scotland

B: Henry VII

C: James V of Scotland

D: Henry VIII

47

Who succeeded William Taft as US president in 1913?

A: Grover Cleveland

B: Theodore Roosevelt

C: Woodrow Wilson

D: Calvin Coolidge

48

In which Surrey town was the classic sitcom 'Terry And June' set?

A: Guildford

B: Purley

C: Dorking

D: Reigate

49

Which Stadium was Specially built for the 1908 London Olympics?

A: Wembley

B: White City

C: Stamford Bridge

D: Highbury

50

Which of these is a middle name of Princess Beatrice of York?

A: Elizabeth

B: Victoria

C: Jane

D: Catherine

If you would like to use your 50:50 please turn to page 275
To ask the audience please turn to page 295
Turn to the answer section on page 306 to find out if you've won £250,000!

14 ◆ £500,000

1

The alphabet used in Russia is named after which saint?

A: Barnabus

B: Justinian

C: Cyril

D: Gregory

2

Which of these actors was born Harvey Yeary?

A: Tom Selleck

B: Lee Majors

C: Don Johnson

D: Martin Sheen

3

Nightblindness is normally the first sign
of a deficiency of which vitamin?

A: A

B: C

C: D

D: E

4

Who gained the rare distinction of taking 4 wickets
with 5 balls in an international cricket match in 1970?

A: Mike Proctor

B: Eddie Barlow

C: Trevor Goddard

D: Garth Le Roux

5

What is the capital of Connecticut?

A: Helena

B: Montpelier

C: Hartford

D: Madison

If you would like to use your 50:50 please turn to page 275
To ask the audience please turn to page 295
Turn to the answer section on page 306 to find out if you've won £500,000!

6

In ancient Greece, what name was given to a public official?

- **A:** Areca
- **B:** Ares
- **C:** Argus
- **D:** Archon

7

What is Wednesday Addams' middle name in 'The Addams Family'?

- **A:** Monday
- **B:** Tuesday
- **C:** Thursday
- **D:** Friday

8

Alfred Mosher Butts is credited with inventing which board game?

- **A:** Monopoly
- **B:** Risk
- **C:** Cluedo
- **D:** Scrabble

9

Where were the 1948 Summer Olympic Games held?

- **A:** London
- **B:** Berlin
- **C:** Moscow
- **D:** Rome

10

What was the first name of Dr Banda, the first president of Malawi?

- **A:** Hastings
- **B:** Dover
- **C:** Worthing
- **D:** Rye

If you would like to use your 50:50 please turn to page 275

To ask the audience please turn to page 295

Turn to the answer section on page 306 to find out if you've won £500,000!

14◆£500,000

11

Which member of the Bach family
was known as the 'English Bach'?

A: Johann Sebastian **B:** Carl Philipp Emanuel

C: Johann Christian **D:** Wilhelm Friedmann

12

Norwegian Knut Hamsun was awarded which Nobel Prize?

A: Medicine **B:** Physics

C: Economics **D:** Literature

13

In German mythology, what is a Nibelung?

A: Dwarf **B:** God

C: Man **D:** Giant

14

How did the millionaire Andrew Carnegie make his money?

A: Steel **B:** Furs

C: Diamonds **D:** Hotels

15

What is the capital of Connecticut?

A: Helena **B:** Montpelier

C: Hartford **D:** Madison

If you would like to use your 50:50 please turn to page 275
To ask the audience please turn to page 295
Turn to the answer section on page 306 to find out if you've won £500,000!

14 ◆ £500,000

16

What nationality was the statesman Le Duc Tho?

- **A:** Burmese
- **B:** Cambodian
- **C:** Korean
- **D:** Vietnamese

17

Who did Derek Jacobi play in the 1996 film of 'Hamlet'?

- **A:** Claudius
- **B:** Polonius
- **C:** Laertes
- **D:** Horatio

18

With which political organisation
is Chief Buthelezi most associated?

- **A:** Inkatha
- **B:** SWAPO
- **C:** ANC
- **D:** Zanu-PF

19

Who were the gang members featured
in the novel 'A Clockwork Orange'?

- **A:** Coogs
- **B:** Troogs
- **C:** Moogs
- **D:** Droogs

20

Ostrava is a city in which European country?

- **A:** Bulgaria
- **B:** Hungary
- **C:** Czech Republic
- **D:** Romania

If you would like to use your 50:50 please turn to page 275
To ask the audience please turn to page 295
Turn to the answer section on page 306 to find out if you've won £500,000!

14 ◆ £500,000

21

Who is the main male character
in 'The Taming of the Shrew'?

A: Angelo

B: Orlando

C: Petruchio

D: Bassanio

22

How was the Colossus of Rhodes destroyed?

A: Earthquake

B: Lightning strike

C: Volcanic eruption

D: Mortar attack

23

In mythology, who killed Hector?

A: Achilles

B: Oedipus

C: Ajax

D: Orestes

24

Which African country is home to the Riff people?

A: Ghana

B: Botswana

C: Chad

D: Morocco

25

In bookmakers' slang, what does 'carpet' mean?

A: Evens

B: 5/4

C: 3/1

D: 500/1

If you would like to use your 50:50 please turn to page 275
To ask the audience please turn to page 295
Turn to the answer section on page 306 to find out if you've won £500,000!

14 ◆ £500,000

26

The Rye House Plot was a conspiracy to assassinate which monarch?

A: Elizabeth I   **B:** Charles II

C: George III   **D:** William IV

27

Which jazz singer is particularly associated with the song 'Strange Fruit'?

A: Ella Fitzgerald   **B:** Nina Simone

C: Billie Holiday   **D:** Eva Cassidy

28

Which language is also known as Lettish?

A: Latvian   **B:** Latin

C: Lapp   **D:** Lithuanian

29

'The Chocolate Soldier' is a musical adaptation of which George Bernard Shaw play?

A: Man and Superman   **B:** Arms and the Man

C: The Applecart   **D:** O'Flaherty VC

30

In the USA, which public holiday is celebrated on the third Monday in February?

A: Thanksgiving Day   **B:** Presidents' Day

C: Martin Luther King Day   **D:** Labor Day

If you would like to use your 50:50 please turn to page 275
To ask the audience please turn to page 295
Turn to the answer section on page 306 to find out if you've won £500,000!

14 ◆ £500,000

31

Saint Augustine, the oldest city in the USA, is in which state?

- **A:** South Carolina
- **B:** Virginia
- **C:** New Hampshire
- **D:** Florida

32

Who was the Babylonian goddess of love and war?

- **A:** Ashnan
- **B:** Mithras
- **C:** Ishtar
- **D:** Aya

33

In which of these cities is there a district called Kings Cross?

- **A:** Montreal
- **B:** Sydney
- **C:** Washington
- **D:** Wellington

34

Which southwest London borough was the coronation place of seven Saxon kings?

- **A:** Richmond
- **B:** Hammersmith & Fulham
- **C:** Kingston
- **D:** Wandsworth

35

Which of the following means 'way of the gods'?

- **A:** Buddhism
- **B:** Sikhism
- **C:** Jainism
- **D:** Shintoism

If you would like to use your 50:50 please turn to page 275

To ask the audience please turn to page 295

Turn to the answer section on page 306 to find out if you've won £500,000!

14 ◆ £500,000

36

What was the name of 'Dangermouse' villain Baron Greenback's fluffy white pet?

A: Augustus

B: Caligula

C: Nero

D: Tiberius

37

Who is not a fairy in Shakespeare's 'A Midsummer Night's Dream'?

A: Cobweb

B: Peaseblossom

C: Mustardseed

D: Parsley

38

What does the name 'Genghis Khan' mean?

A: Faster rider

B: Mighty Swordsman

C: Ruler of all

D: Father of many

39

'The Tortellis' was a little-known spin-off of which US sitcom?

A: Soap

B: Cheers

C: Frasier

D: Happy Days

40

Talisker whisky is distilled on which island?

A: Lewis

B: Islay

C: Mull

D: Skye

If you would like to use your 50:50 please turn to page 275
To ask the audience please turn to page 295
Turn to the answer section on page 307 to find out if you've won £500,000!

243

14 ◆ £500,000

41

Who first coined the expression 'ladies who lunch'?

A: Noël Coward

B: Stephen Sondheim

C: W.S. Gilbert

D: Oscar Wilde

42

Which bird's plumage turns white in winter?

A: Kestrel

B: Ptarmigan

C: Snowy owl

D: Corn bunting

43

When Victoria ascended the throne in 1837, which relation did she succeed?

A: Father

B: Brother

C: Uncle

D: Cousin

44

Which Nobel Prize-winning author wrote 'Herzog' and 'Humboldt's Gift'?

A: Joseph Brodsky

B: John Steinbeck

C: Saul Bellow

D: Samuel Becket

45

Of which group of islands is Muckle Flugga a part?

A: Isles of Scilly

B: Inner Hebrides

C: Orkney Islands

D: Shetland Islands

If you would like to use your 50:50 please turn to page 275
To ask the audience please turn to page 295
Turn to the answer section on page 307 to find out if you've won £500,000!

14 ◆ £500,000

46

In which US city was the UN Charter drawn up in 1945?

◆ **A:** San Francisco   ◆ **B:** Seattle

◆ **C:** Philadelphia   ◆ **D:** Washington

47

In which country did the so-called
'Carnation Revolution' take place in 1974?

◆ **A:** Portugal   ◆ **B:** Spain

◆ **C:** Mexico   ◆ **D:** Brazil

48

Which of the following is a version of the card game rummy?

◆ **A:** Chicago   ◆ **B:** Oklahoma

◆ **C:** Carousel   ◆ **D:** Copacabana

49

Of which of the following is
Saint Nicholas not the patron saint?

◆ **A:** Russia   ◆ **B:** Aberdeen

◆ **C:** Pawnbrokers   ◆ **D:** Watchmakers

50

In which country are so-called 'lucky grapes'
traditionally eaten on New Year's Eve?

◆ **A:** Italy   ◆ **B:** Germany

◆ **C:** Spain   ◆ **D:** France

If you would like to use your 50:50 please turn to page 276
To ask the audience please turn to page 296
Turn to the answer section on page 307 to find out if you've won £500,000!

15 ◆ £1,000,000

1

In 1985, Clive Ponting was charged with leaking details of the sinking of the Belgrano to whom?

A: Dennis Skinner   **B:** Tam Dalyell

C: Menzies Campbell   **D:** George Galloway

2

How is ortho-sulfobenzoic acid imide more commonly known?

A: Fructose   **B:** Quinine

C: Caffeine   **D:** Saccharin

3

Which member of the Royal Family has the forename Louise?

A: Princess Royal   **B:** Countess of Wessex

C: Duchess of Kent   **D:** Princess Eugenie

4

With what majority did the Conservatives surprisingly win the 1970 General Election?

A: 5   **B:** 18

C: 30   **D:** 47

5

Which of these British universities is the oldest?

A: Cambridge   **B:** Exeter

C: Reading   **D:** Birmingham

If you would like to use your 50:50 please turn to page 276
To ask the audience please turn to page 296
Turn to the answer section on page 307 to find out if you've won £1,000,000!

15 ◆ £1,000,000

6

In which country was the world's first air-mail service initiated in 1911?

◆ **A:** Australia   ◆ **B:** USA

◆ **C:** France   ◆ **D:** UK

7

What was the profession of the composer Borodin?

◆ **A:** Naval captain   ◆ **B:** Chemist

◆ **C:** Lawyer   ◆ **D:** Chef

8

Who of the following was not one of Hibernian's 'Famous Five' forward line?

◆ **A:** Gordon Smith   ◆ **B:** Bobby Johnstone

◆ **C:** Willie Ormond   ◆ **D:** Tommy Younger

9

What was the original name of Cliff Richard's backing band 'The Shadows'?

◆ **A:** The Miracles   ◆ **B:** The Drifters

◆ **C:** The Crickets   ◆ **D:** The Phantoms

10

What is the penultimate book of the Old Testament?

◆ **A:** Habbakuk   ◆ **B:** Micah

◆ **C:** Obadiah   ◆ **D:** Zechariah

If you would like to use your 50:50 please turn to page 276

To ask the audience please turn to page 296

Turn to the answer section on page 307 to find out if you've won £1,000,000!

11

Which fruit is most likely to
suffer from the disease 'big bud'?

A: Blackcurrant **B:** Mango

C: Melon **D:** Kumquat

12

On a ship, what is a 'davit'?

A: Sail **B:** Crane

C: Anchor **D:** Compass

13

What nationality was the composer Smetana?

A: Hungarian **B:** Russian

C: Czech **D:** French

14

Darius the Great was king of which ancient country?

A: India **B:** Egypt

C: Japan **D:** Persia

15

The site of the city of Nineveh is in which country?

A: Jordan **B:** Israel

C: Iran **D:** Iraq

If you would like to use your 50:50 please turn to page 276
To ask the audience please turn to page 296
Turn to the answer section on page 307 to find out if you've won £1,000,000!

16

In the 18th century, Thomas Tompion was known as a maker of what?

- **A:** Guns
- **B:** Clocks
- **C:** Wigs
- **D:** Furniture

17

Pastinaca sativa is the Latin name for which vegetable?

- **A:** Carrot
- **B:** Potato
- **C:** Cabbage
- **D:** Parsnip

18

At what sustained wind speed are tropical storms upgraded to hurricanes?

- **A:** 52 mph
- **B:** 60 mph
- **C:** 74 mph
- **D:** 88 mph

19

The Bellingshausen Sea is part of which ocean?

- **A:** Atlantic
- **B:** Pacific
- **C:** Indian
- **D:** Arctic

20

In which year was 'Merry Xmas Everybody' a Christmas UK No 1 hit for Slade?

- **A:** 1973
- **B:** 1972
- **C:** 1975
- **D:** 1974

If you would like to use your 50:50 please turn to page 276
To ask the audience please turn to page 296
Turn to the answer section on page 307 to find out if you've won £1,000,000!

15◆£1,000,000

21
'Cuneiform' is a type of what?

A: Dancing   **B:** Music

C: Medicine   **D:** Writing

22
Who was the husband of Lady Godiva?

A: Beowulf   **B:** Aethelbert

C: Godwin   **D:** Leofric

23
Which extinct creature was also known as a 'thylacine'?

A: Moa   **B:** Ground sloth

C: Mastodon   **D:** Tasmanian wolf

24
The word 'slalom' comes from which language?

A: Gaelic   **B:** Norwegian

C: Russian   **D:** Finnish

25
Connected with witches, what is a 'cantrip'?

A: Black cat   **B:** Broomstick

C: Pointed hat   **D:** Magic spell

If you would like to use your 50:50 please turn to page 276
To ask the audience please turn to page 296
Turn to the answer section on page 307 to find out if you've won £1,000,000!

26

Which of these people would use 'noble rot' in their work?

- **A:** Barrelmaker
- **B:** Cheesemaker
- **C:** Lacemaker
- **D:** Winemaker

27

Which prime minister repealed the Corn Laws?

- **A:** Peel
- **B:** Gladstone
- **C:** Wellington
- **D:** Pitt

28

Which tennis player appeared in the video of 'Escape' by Enrique Iglesias?

- **A:** Venus Williams
- **B:** Jennifer Capriati
- **C:** Martina Hingis
- **D:** Anna Kournikova

29

Which three initials comprised the distress call prior to SOS, which was made standard in 1908?

- **A:** EME
- **B:** CQD
- **C:** SID
- **D:** IDD

30

Sauchiehall Street is a major shopping street of which city?

- **A:** Newcastle
- **B:** Glasgow
- **C:** Bristol
- **D:** Manchester

If you would like to use your 50:50 please turn to page 276
To ask the audience please turn to page 296
Turn to the answer section on page 307 to find out if you've won £1,000,000!

15 ◆ £1,000,000

31

What is the name of the musical, premiered in 2002, based on the songs of the rock band Queen?

A: We Will Rock You

B: Bohemian Rhapsody

C: Somebody to Love

D: Under Pressure

32

In the Austin Powers films, what is the name of Dr Evil's small sidekick?

A: Deary-Me

B: Little-Me

C: Mini-Me

D: Me-Me

33

A number one followed by one hundred zeros is known by what name?

A: Googol

B: Megatron

C: Gigabit

D: Nanomole

34

In Japanese cuisine 'mushimono' refers to food which has been . . .?

A: Baked

B: Puréed

C: Deep fried

D: Steamed

35

'I'm all there with me cough drops' was a catchphrase of which popular comedian?

A: Al Read

B: Ted Ray

C: Jimmy Clitheroe

D: Kenneth Horne

If you would like to use your 50:50 please turn to page 276
To ask the audience please turn to page 296
Turn to the answer section on page 307 to find out if you've won £1,000,000!

36

Which classic British sitcom evolved from
a play called 'The Banana Box'?

A: Porridge

B: Open all Hours

C: Rising Damp

D: Dad's Army

37

Which line, proposed in 1919, divided Russia and Poland?

A: Alba Line

B: Attila Line

C: Balfour Line

D: Curzon Line

38

Of which UK comedy show was
'Les Guignols' a French adaptation?

A: 'Allo 'Allo!

B: The Young Ones

C: Spitting Image

D: Dad's Army

39

1972 saw the first appearance of
which of these quiz shows on UK television?

A: 3-2-1

B: Runaround

C: Mastermind

D: Mr & Mrs

40

Which Apollo 13 astronaut unwittingly
coined a catchphrase in 1970?

A: James Lovell

B: John Swigert Jr

C: Fred Haise Jr

D: Al Shephard

If you would like to use your 50:50 please turn to page 276
To ask the audience please turn to page 296
Turn to the answer section on page 307 to find out if you've won £1,000,000!

15◆£1,000,000

41

Which of these was a musical effects device invented by Godley and Creme?

A: The Bobbit

B: The Gizmo

C: The Doodah

D: The Whatsit

42

Who performs the rap song 'Christmas in Hollis', featured in the film 'Die Hard'?

A: Grandmaster Flash

B: Run DMC

C: LL Cool J

D: Schoolly D

43

What did the engineer George Stephenson call his first locomotive?

A: Rocket

B: The Salamanca

C: Puffing Billy

D: Blücher

44

Who graduated from Oxford University at the tender age of thirteen in 1985?

A: Naomi Stevens

B: Ruth Lawrence

C: Miriam Thierens

D: Deborah Witchell

45

How many Top 20 UK singles did the group Madness have during the 1980s?

A: 15

B: 17

C: 19

D: 21

If you would like to use your 50:50 please turn to page 276
To ask the audience please turn to page 296
Turn to the answer section on page 307 to find out if you've won £1,000,000!

15♦£1,000,000

46

Which of the following was the recipient
of the Nobel Prize for Peace in 1989?

A: Dalai Lama

B: Pope John Paul II

C: Mikhail Gorbachev

D: United Nations

47

Who famously said of Ingrid Bergman that
'she speaks 5 languages and can't act in any of them'?

A: John Gielgud

B: Laurence Olivier

C: Alec Guinness

D: Charles Hawtrey

48

Which 20th-century designer used a moulding
machine nicknamed the 'Kazam!'?

A: Christopher Dresser

B: Charles Rennie Mackintosh

C: Charles Eames

D: Robin Day

49

Which Roman ruler is described in sources as ordering his troops
to 'gather seashells' after abandoning an invasion of Britain?

A: Nero

B: Caligula

C: Claudius

D: Tiberius

50

From which Italian city are the narrators
of the tales in Boccaccio's 'Decameron'?

A: Rome

B: Milan

C: Florence

D: Lucca

If you would like to use your 50:50 please turn to page 276
To ask the audience please turn to page 296
Turn to the answer section on page 307 to find out if you've won £1,000,000!

50:50

£100

1	options remaining are A & D	30	options remaining are A & C
2	options remaining are A & C	31	options remaining are A & C
3	options remaining are B & D	32	options remaining are A & B
4	options remaining are A & D	33	options remaining are C & D
5	options remaining are A & C	34	options remaining are A & C
6	options remaining are C & D	35	options remaining are A & B
7	options remaining are A & C	36	options remaining are A & D
8	options remaining are A & D	37	options remaining are A & B
9	options remaining are B & D	38	options remaining are B & C
10	options remaining are A & B	39	options remaining are A & C
11	options remaining are B & C	40	options remaining are A & D
12	options remaining are A & C	41	options remaining are C & D
13	options remaining are B & D	42	options remaining are A & C
14	options remaining are A & D	43	options remaining are B & C
15	options remaining are B & C	44	options remaining are A & B
16	options remaining are A & C	45	options remaining are B & C
17	options remaining are B & C	46	options remaining are A & B
18	options remaining are A & B	47	options remaining are A & C
19	options remaining are A & C	48	options remaining are B & C
20	options remaining are A & B	49	options remaining are A & B
21	options remaining are A & D	50	options remaining are A & D
22	options remaining are A & C	51	options remaining are B & C
23	options remaining are B & C	52	options remaining are A & B
24	options remaining are A & B	53	options remaining are B & C
25	options remaining are A & C	54	options remaining are A & B
26	options remaining are A & D	55	options remaining are A & C
27	options remaining are A & B	56	options remaining are A & D
28	options remaining are A & B	57	options remaining are C & D
29	options remaining are B & D	58	options remaining are A & B

50:50

£100

59 options remaining are A & C	88 options remaining are A & C
60 options remaining are A & B	89 options remaining are A & B
61 options remaining are A & D	90 options remaining are A & D
62 options remaining are A & B	91 options remaining are B & C
63 options remaining are B & C	92 options remaining are A & C
64 options remaining are B & C	93 options remaining are B & D
65 options remaining are A & B	94 options remaining are A & B
66 options remaining are B & D	95 options remaining are A & D
67 options remaining are B & C	96 options remaining are C & D
68 options remaining are A & C	97 options remaining are B & C
69 options remaining are A & C	98 options remaining are A & C
70 options remaining are C & D	99 options remaining are B & D
71 options remaining are A & C	100 options remaining are A & C
72 options remaining are A & D	

£200

73 options remaining are C & D	1 options remaining are A & C
74 options remaining are B & C	2 options remaining are C & D
75 options remaining are B & D	3 options remaining are C & D
76 options remaining are B & C	4 options remaining are A & D
77 options remaining are B & D	5 options remaining are B & C
78 options remaining are A & D	6 options remaining are A & C
79 options remaining are A & C	7 options remaining are B & C
80 options remaining are B & C	8 options remaining are C & D
81 options remaining are A & C	9 options remaining are A & C
82 options remaining are A & B	10 options remaining are B & C
83 options remaining are B & D	11 options remaining are B & C
84 options remaining are C & D	12 options remaining are B & D
85 options remaining are B & C	13 options remaining are A & D
86 options remaining are A & B	14 options remaining are C & D
87 options remaining are A & C	

50:50

£200

15 options remaining are A & B	44 options remaining are B & C
16 options remaining are A & C	45 options remaining are A & C
17 options remaining are C & D	46 options remaining are A & B
18 options remaining are B & D	47 options remaining are A & D
19 options remaining are A & C	48 options remaining are A & C
20 options remaining are C & D	49 options remaining are B & C
21 options remaining are A & C	50 options remaining are A & D
22 options remaining are C & D	51 options remaining are A & B
23 options remaining are A & B	52 options remaining are B & C
24 options remaining are B & C	53 options remaining are B & D
25 options remaining are A & D	54 options remaining are A & D
26 options remaining are A & B	55 options remaining are B & C
27 options remaining are B & C	56 options remaining are A & B
28 options remaining are A & D	57 options remaining are A & C
29 options remaining are A & B	58 options remaining are B & C
30 options remaining are C & D	59 options remaining are A & D
31 options remaining are A & D	60 options remaining are A & B
32 options remaining are B & C	61 options remaining are A & D
33 options remaining are A & D	62 options remaining are C & D
34 options remaining are B & C	63 options remaining are B & C
35 options remaining are A & B	64 options remaining are A & D
36 options remaining are A & C	65 options remaining are A & C
37 options remaining are B & D	66 options remaining are B & D
38 options remaining are C & D	67 options remaining are A & B
39 options remaining are A & D	68 options remaining are B & D
40 options remaining are B & C	69 options remaining are A & C
41 options remaining are C & D	70 options remaining are B & C
42 options remaining are A & B	71 options remaining are A & C
43 options remaining are A & B	72 options remaining are B & D

50:50

£200

73 options remaining are B & C
74 options remaining are A & D
75 options remaining are B & C
76 options remaining are A & D
77 options remaining are A & C
78 options remaining are C & D
79 options remaining are A & C
80 options remaining are B & D
81 options remaining are A & C
82 options remaining are A & B
83 options remaining are A & D
84 options remaining are C & D
85 options remaining are B & C
86 options remaining are B & D
87 options remaining are A & B
88 options remaining are B & C
89 options remaining are A & C
90 options remaining are A & D
91 options remaining are A & C
92 options remaining are B & D
93 options remaining are A & C
94 options remaining are A & D
95 options remaining are A & B
96 options remaining are A & D
97 options remaining are A & B
98 options remaining are B & C
99 options remaining are B & D
100 options remaining are A & B

£300

1 options remaining are B & C
2 options remaining are A & B
3 options remaining are B & C
4 options remaining are A & D
5 options remaining are B & C
6 options remaining are B & C
7 options remaining are A & C
8 options remaining are C & D
9 options remaining are B & C
10 options remaining are A & B
11 options remaining are B & C
12 options remaining are C & D
13 options remaining are B & D
14 options remaining are B & D
15 options remaining are A & C
16 options remaining are B & C
17 options remaining are C & D
18 options remaining are A & B
19 options remaining are C & D
20 options remaining are A & C
21 options remaining are C & D
22 options remaining are A & C
23 options remaining are B & D
24 options remaining are B & C
25 options remaining are A & D
26 options remaining are B & C
27 options remaining are A & D
28 options remaining are B & C

£300

29	options remaining are A & C	57	options remaining are B & C
30	options remaining are C & D	58	options remaining are B & D
31	options remaining are B & D	59	options remaining are A & B
32	options remaining are A & C	60	options remaining are B & C
33	options remaining are A & D	61	options remaining are A & B
34	options remaining are C & D	62	options remaining are B & D
35	options remaining are B & D	63	options remaining are B & C
36	options remaining are A & C	64	options remaining are A & D
37	options remaining are B & C	65	options remaining are B & C
38	options remaining are C & D	66	options remaining are B & D
39	options remaining are B & D	67	options remaining are A & D
40	options remaining are A & C	68	options remaining are B & C
41	options remaining are C & D	69	options remaining are A & C
42	options remaining are A & B	70	options remaining are A & B
43	options remaining are A & C	71	options remaining are B & C
44	options remaining are B & D	72	options remaining are A & C
45	options remaining are A & D	73	options remaining are A & B
46	options remaining are B & C	74	options remaining are C & D
47	options remaining are A & B	75	options remaining are A & D
48	options remaining are A & C	76	options remaining are A & B
49	options remaining are B & D	77	options remaining are B & D
50	options remaining are B & C	78	options remaining are A & B
51	options remaining are C & D	79	options remaining are A & C
52	options remaining are A & B	80	options remaining are B & D
53	options remaining are B & D	81	options remaining are A & C
54	options remaining are A & C	82	options remaining are B & C
55	options remaining are B & C	83	options remaining are B & C
56	options remaining are A & D	84	options remaining are A & D

50:50

£300

85 options remaining are A & B
86 options remaining are A & C
87 options remaining are B & C
88 options remaining are C & D
89 options remaining are B & D
90 options remaining are B & C
91 options remaining are A & B
92 options remaining are A & D
93 options remaining are B & C
94 options remaining are C & D
95 options remaining are A & C
96 options remaining are B & D
97 options remaining are A & B
98 options remaining are B & C
99 options remaining are C & D
100 options remaining are B & C

£500

1 options remaining are A & B
2 options remaining are A & C
3 options remaining are C & D
4 options remaining are B & D
5 options remaining are B & C
6 options remaining are B & D
7 options remaining are A & C
8 options remaining are A & D
9 options remaining are B & C
10 options remaining are A & B
11 options remaining are B & C

12 options remaining are A & D
13 options remaining are C & D
14 options remaining are B & D
15 options remaining are A & C
16 options remaining are B & D
17 options remaining are A & C
18 options remaining are B & D
19 options remaining are B & C
20 options remaining are A & C
21 options remaining are B & C
22 options remaining are A & D
23 options remaining are B & C
24 options remaining are A & D
25 options remaining are B & C
26 options remaining are B & C
27 options remaining are A & B
28 options remaining are C & D
29 options remaining are B & D
30 options remaining are C & D
31 options remaining are A & B
32 options remaining are A & D
33 options remaining are B & D
34 options remaining are A & B
35 options remaining are B & C
36 options remaining are B & C
37 options remaining are B & D
38 options remaining are A & B
39 options remaining are C & D
40 options remaining are A & C

50:50

£500

41 options remaining are B & C
42 options remaining are B & D
43 options remaining are A & C
44 options remaining are A & C
45 options remaining are C & D
46 options remaining are C & D
47 options remaining are B & C
48 options remaining are A & C
49 options remaining are A & D
50 options remaining are B & C
51 options remaining are A & C
52 options remaining are B & D
53 options remaining are C & D
54 options remaining are A & B
55 options remaining are B & D
56 options remaining are A & C
57 options remaining are A & D
58 options remaining are B & C
59 options remaining are A & B
60 options remaining are A & D
61 options remaining are C & D
62 options remaining are B & C
63 options remaining are A & D
64 options remaining are B & C
65 options remaining are A & D
66 options remaining are B & C
67 options remaining are B & D
68 options remaining are A & B
69 options remaining are C & D

70 options remaining are A & C
71 options remaining are C & D
72 options remaining are B & C
73 options remaining are A & D
74 options remaining are C & D
75 options remaining are A & D
76 options remaining are A & B
77 options remaining are C & D
78 options remaining are A & B
79 options remaining are B & D
80 options remaining are A & B
81 options remaining are B & C
82 options remaining are A & C
83 options remaining are A & D
84 options remaining are B & C
85 options remaining are C & D
86 options remaining are A & C
87 options remaining are A & B
88 options remaining are A & C
89 options remaining are B & C
90 options remaining are C & D
91 options remaining are B & C
92 options remaining are A & B
93 options remaining are B & C
94 options remaining are B & D
95 options remaining are A & C
96 options remaining are B & C
97 options remaining are B & D
98 options remaining are A & D

50:50

£500

99 options remaining are C & D
100 options remaining are A & D

£1,000

1 options remaining are A & B
2 options remaining are C & D
3 options remaining are A & D
4 options remaining are B & C
5 options remaining are C & D
6 options remaining are B & C
7 options remaining are C & D
8 options remaining are A & D
9 options remaining are A & B
10 options remaining are B & D
11 options remaining are A & D
12 options remaining are A & B
13 options remaining are A & D
14 options remaining are C & D
15 options remaining are B & C
16 options remaining are A & D
17 options remaining are B & C
18 options remaining are A & D
19 options remaining are B & C
20 options remaining are C & D
21 options remaining are A & B
22 options remaining are A & C
23 options remaining are C & D
24 options remaining are C & D
25 options remaining are A & D

26 options remaining are B & D
27 options remaining are A & C
28 options remaining are B & C
29 options remaining are A & D
30 options remaining are B & D
31 options remaining are A & B
32 options remaining are B & C
33 options remaining are B & D
34 options remaining are B & C
35 options remaining are B & D
36 options remaining are A & B
37 options remaining are B & C
38 options remaining are A & B
39 options remaining are A & C
40 options remaining are B & C
41 options remaining are A & C
42 options remaining are A & D
43 options remaining are A & C
44 options remaining are A & B
45 options remaining are B & C
46 options remaining are C & D
47 options remaining are B & D
48 options remaining are A & C
49 options remaining are C & D
50 options remaining are A & B
51 options remaining are C & D
52 options remaining are A & C
53 options remaining are B & C
54 options remaining are A & B

50:50

£1,000

55 options remaining are C & D
56 options remaining are B & D
57 options remaining are A & B
58 options remaining are A & C
59 options remaining are B & D
60 options remaining are A & B
61 options remaining are B & C
62 options remaining are A & D
63 options remaining are C & D
64 options remaining are A & B
65 options remaining are C & D
66 options remaining are A & C
67 options remaining are C & D
68 options remaining are A & C
69 options remaining are A & B
70 options remaining are B & C
71 options remaining are A & C
72 options remaining are B & C
73 options remaining are A & B
74 options remaining are B & D
75 options remaining are B & C
76 options remaining are A & D
77 options remaining are B & C
78 options remaining are C & D
79 options remaining are A & B
80 options remaining are B & D
81 options remaining are C & D
82 options remaining are A & C
83 options remaining are A & B

84 options remaining are C & D
85 options remaining are B & C
86 options remaining are B & C
87 options remaining are A & B
88 options remaining are C & D
89 options remaining are A & B
90 options remaining are B & C
91 options remaining are B & C
92 options remaining are A & B
93 options remaining are A & D
94 options remaining are B & C
95 options remaining are B & C
96 options remaining are A & C
97 options remaining are B & C
98 options remaining are A & D
99 options remaining are B & C
100 options remaining are A & D

£2,000

1 options remaining are B & C
2 options remaining are A & B
3 options remaining are C & D
4 options remaining are A & D
5 options remaining are C & D
6 options remaining are B & C
7 options remaining are A & D
8 options remaining are A & C
9 options remaining are A & D
10 options remaining are A & B

50:50

£2,000

11 options remaining are C & D	40 options remaining are B & D
12 options remaining are A & D	41 options remaining are A & B
13 options remaining are A & B	42 options remaining are B & C
14 options remaining are B & C	43 options remaining are A & D
15 options remaining are C & D	44 options remaining are B & D
16 options remaining are A & B	45 options remaining are A & B
17 options remaining are C & D	46 options remaining are A & C
18 options remaining are A & C	47 options remaining are A & D
19 options remaining are A & B	48 options remaining are B & D
20 options remaining are A & C	49 options remaining are A & C
21 options remaining are B & C	50 options remaining are A & B
22 options remaining are A & B	51 options remaining are B & C
23 options remaining are A & B	52 options remaining are B & D
24 options remaining are B & C	53 options remaining are C & D
25 options remaining are C & D	54 options remaining are B & C
26 options remaining are B & C	55 options remaining are A & C
27 options remaining are A & C	56 options remaining are A & D
28 options remaining are B & D	57 options remaining are B & C
29 options remaining are C & D	58 options remaining are B & C
30 options remaining are A & D	59 options remaining are B & D
31 options remaining are B & C	60 options remaining are A & C
32 options remaining are B & D	61 options remaining are A & D
33 options remaining are A & D	62 options remaining are A & D
34 options remaining are B & D	63 options remaining are B & C
35 options remaining are A & C	64 options remaining are B & C
36 options remaining are A & D	65 options remaining are A & C
37 options remaining are C & D	66 options remaining are B & C
38 options remaining are A & D	67 options remaining are C & D
39 options remaining are B & C	68 options remaining are A & B

50:50

£2,000

69 options remaining are A & C	21 options remaining are B & C
70 options remaining are C & D	22 options remaining are B & D
71 options remaining are C & D	23 options remaining are A & C
72 options remaining are A & C	24 options remaining are C & D
73 options remaining are B & C	25 options remaining are B & C
74 options remaining are B & D	26 options remaining are B & C
75 options remaining are A & B	27 options remaining are B & C
	28 options remaining are B & D
£4,000	29 options remaining are B & D
1 options remaining are A & B	30 options remaining are A & B
2 options remaining are C & D	31 options remaining are C & D
3 options remaining are A & B	32 options remaining are A & C
4 options remaining are B & C	33 options remaining are A & D
5 options remaining are C & D	34 options remaining are B & D
6 options remaining are A & C	35 options remaining are B & C
7 options remaining are C & D	36 options remaining are C & D
8 options remaining are A & B	37 options remaining are B & C
9 options remaining are B & D	38 options remaining are C & D
10 options remaining are A & C	39 options remaining are B & D
11 options remaining are B & D	40 options remaining are B & C
12 options remaining are C & D	41 options remaining are A & B
13 options remaining are A & B	42 options remaining are B & C
14 options remaining are A & C	43 options remaining are C & D
15 options remaining are A & B	44 options remaining are B & C
16 options remaining are B & C	45 options remaining are C & D
17 options remaining are B & C	46 options remaining are A & D
18 options remaining are C & D	47 options remaining are B & C
19 options remaining are A & C	48 options remaining are A & D
20 options remaining are B & D	49 options remaining are B & D

50:50

£4,000

50 options remaining are C & D	2 options remaining are A & C
51 options remaining are C & D	3 options remaining are C & D
52 options remaining are B & C	4 options remaining are A & B
53 options remaining are B & C	5 options remaining are B & C
54 options remaining are C & D	6 options remaining are C & D
55 options remaining are A & B	7 options remaining are A & C
56 options remaining are B & C	8 options remaining are B & C
57 options remaining are B & D	9 options remaining are A & D
58 options remaining are B & C	10 options remaining are B & D
59 options remaining are A & B	11 options remaining are B & C
60 options remaining are C & D	12 options remaining are A & D
61 options remaining are A & D	13 options remaining are A & B
62 options remaining are C & D	14 options remaining are A & C
63 options remaining are A & D	15 options remaining are C & D
64 options remaining are B & D	16 options remaining are A & B
65 options remaining are A & C	17 options remaining are B & D
66 options remaining are C & D	18 options remaining are B & C
67 options remaining are A & D	19 options remaining are A & B
68 options remaining are B & C	20 options remaining are B & D
69 options remaining are B & D	21 options remaining are A & D
70 options remaining are A & B	22 options remaining are A & C
71 options remaining are A & C	23 options remaining are B & C
72 options remaining are A & B	24 options remaining are C & D
73 options remaining are C & D	25 options remaining are B & D
74 options remaining are B & D	26 options remaining are B & C
75 options remaining are C & D	27 options remaining are A & B
	28 options remaining are B & C
£8,000	29 options remaining are A & D
1 options remaining are C & D	30 options remaining are C & D

50:50

£8,000

31 options remaining are B & C
32 options remaining are A & B
33 options remaining are C & D
34 options remaining are B & D
35 options remaining are A & C
36 options remaining are B & C
37 options remaining are A & C
38 options remaining are B & D
39 options remaining are A & B
40 options remaining are C & D
41 options remaining are A & C
42 options remaining are B & D
43 options remaining are A & C
44 options remaining are A & B
45 options remaining are A & C
46 options remaining are A & D
47 options remaining are B & D
48 options remaining are A & C
49 options remaining are B & D
50 options remaining are A & C
51 options remaining are C & D
52 options remaining are A & C
53 options remaining are B & D
54 options remaining are A & B
55 options remaining are A & C
56 options remaining are B & D
57 options remaining are C & D
58 options remaining are A & D
59 options remaining are C & D

60 options remaining are B & C
61 options remaining are C & D
62 options remaining are A & C
63 options remaining are A & B
64 options remaining are A & C
65 options remaining are C & D
66 options remaining are A & D
67 options remaining are B & C
68 options remaining are A & D
69 options remaining are C & D
70 options remaining are B & D
71 options remaining are B & C
72 options remaining are A & C
73 options remaining are A & D
74 options remaining are A & B
75 options remaining are B & C

£16,000

1 options remaining are B & C
2 options remaining are C & D
3 options remaining are A & C
4 options remaining are B & C
5 options remaining are A & C
6 options remaining are B & D
7 options remaining are A & B
8 options remaining are B & C
9 options remaining are A & D
10 options remaining are C & D
11 options remaining are B & C

50:50

£16,000

12 options remaining are A & D	41 options remaining are A & D
13 options remaining are B & D	42 options remaining are B & C
14 options remaining are B & C	43 options remaining are A & D
15 options remaining are A & B	44 options remaining are B & C
16 options remaining are C & D	45 options remaining are C & D
17 options remaining are A & C	46 options remaining are A & B
18 options remaining are A & B	47 options remaining are B & C
19 options remaining are A & D	48 options remaining are A & D
20 options remaining are B & C	49 options remaining are A & C
21 options remaining are B & D	50 options remaining are B & C
22 options remaining are A & C	51 options remaining are B & D
23 options remaining are A & B	52 options remaining are A & B
24 options remaining are B & D	53 options remaining are C & D
25 options remaining are C & D	54 options remaining are A & B
26 options remaining are C & D	55 options remaining are B & C
27 options remaining are A & C	56 options remaining are A & D
28 options remaining are B & D	57 options remaining are B & C
29 options remaining are A & C	58 options remaining are B & C
30 options remaining are A & D	59 options remaining are C & D
31 options remaining are B & C	60 options remaining are B & C
32 options remaining are A & D	61 options remaining are A & D
33 options remaining are A & C	62 options remaining are B & C
34 options remaining are A & D	63 options remaining are C & D
35 options remaining are B & D	64 options remaining are B & C
36 options remaining are C & D	65 options remaining are C & D
37 options remaining are A & D	66 options remaining are A & D
38 options remaining are B & D	67 options remaining are B & C
39 options remaining are C & D	68 options remaining are B & D
40 options remaining are A & C	69 options remaining are A & C

50:50

£16,000

70 options remaining are A & B
71 options remaining are C & D
72 options remaining are B & C
73 options remaining are A & D
74 options remaining are B & C
75 options remaining are C & D

£32,000

1   options remaining are C & D
2   options remaining are A & D
3   options remaining are A & B
4   options remaining are B & C
5   options remaining are B & D
6   options remaining are A & C
7   options remaining are B & C
8   options remaining are A & D
9   options remaining are A & B
10 options remaining are A & C
11 options remaining are B & D
12 options remaining are A & C
13 options remaining are C & D
14 options remaining are B & C
15 options remaining are A & B
16 options remaining are B & C
17 options remaining are A & D
18 options remaining are B & D
19 options remaining are A & C
20 options remaining are B & C
21 options remaining are A & C

22 options remaining are B & C
23 options remaining are B & D
24 options remaining are A & C
25 options remaining are B & C
26 options remaining are A & B
27 options remaining are B & C
28 options remaining are B & D
29 options remaining are B & C
30 options remaining are A & D
31 options remaining are B & D
32 options remaining are C & D
33 options remaining are A & B
34 options remaining are C & D
35 options remaining are A & C
36 options remaining are B & D
37 options remaining are A & C
38 options remaining are B & C
39 options remaining are C & D
40 options remaining are B & C
41 options remaining are C & D
42 options remaining are A & C
43 options remaining are B & C
44 options remaining are A & D
45 options remaining are B & C
46 options remaining are A & B
47 options remaining are A & B
48 options remaining are B & C
49 options remaining are B & D
50 options remaining are A & B

50:50

£32,000

51	options remaining are B & D
52	options remaining are A & C
53	options remaining are A & D
54	options remaining are B & C
55	options remaining are A & B
56	options remaining are C & D
57	options remaining are A & D
58	options remaining are B & D
59	options remaining are C & D
60	options remaining are A & D
61	options remaining are A & C
62	options remaining are C & D
63	options remaining are A & B
64	options remaining are A & C
65	options remaining are A & D
66	options remaining are B & D
67	options remaining are B & C
68	options remaining are A & B
69	options remaining are C & D
70	options remaining are A & B
71	options remaining are C & D
72	options remaining are A & C
73	options remaining are A & C
74	options remaining are C & D
75	options remaining are B & C

£64,000

1	options remaining are C & D
2	options remaining are A & B
3	options remaining are A & C
4	options remaining are A & D
5	options remaining are B & C
6	options remaining are B & D
7	options remaining are A & D
8	options remaining are B & C
9	options remaining are B & D
10	options remaining are A & D
11	options remaining are C & D
12	options remaining are A & B
13	options remaining are C & D
14	options remaining are B & C
15	options remaining are C & D
16	options remaining are B & C
17	options remaining are A & D
18	options remaining are A & C
19	options remaining are C & D
20	options remaining are B & C
21	options remaining are A & B
22	options remaining are A & C
23	options remaining are B & C
24	options remaining are B & D
25	options remaining are A & B
26	options remaining are A & C
27	options remaining are A & D
28	options remaining are B & C
29	options remaining are A & D
30	options remaining are B & C
31	options remaining are B & D

50:50

£64,000

32	options remaining are A & D	9	options remaining are B & C
33	options remaining are B & C	10	options remaining are B & D
34	options remaining are A & D	11	options remaining are A & C
35	options remaining are C & D	12	options remaining are C & D
36	options remaining are B & D	13	options remaining are B & C
37	options remaining are A & B	14	options remaining are A & C
38	options remaining are C & D	15	options remaining are C & D
39	options remaining are A & B	16	options remaining are A & B
40	options remaining are B & C	17	options remaining are A & B
41	options remaining are C & D	18	options remaining are A & D
42	options remaining are B & D	19	options remaining are B & C
43	options remaining are A & D	20	options remaining are C & D
44	options remaining are C & D	21	options remaining are A & B
45	options remaining are B & C	22	options remaining are A & D
46	options remaining are B & D	23	options remaining are A & B
47	options remaining are A & B	24	options remaining are B & C
48	options remaining are C & D	25	options remaining are C & D
49	options remaining are B & D	26	options remaining are A & C
50	options remaining are A & C	27	options remaining are B & C
		28	options remaining are A & D

£125,000

		29	options remaining are B & C
1	options remaining are B & C	30	options remaining are B & D
2	options remaining are A & C	31	options remaining are B & C
3	options remaining are C & D	32	options remaining are A & D
4	options remaining are B & C	33	options remaining are B & C
5	options remaining are A & C	34	options remaining are A & C
6	options remaining are C & D	35	options remaining are A & B
7	options remaining are A & C	36	options remaining are C & D
8	options remaining are A & B	37	options remaining are A & D

50:50

£125,000

38 options remaining are B & D	14 options remaining are A & D
39 options remaining are A & C	15 options remaining are A & B
40 options remaining are B & C	16 options remaining are B & D
41 options remaining are A & B	17 options remaining are B & C
42 options remaining are A & C	18 options remaining are B & C
43 options remaining are C & D	19 options remaining are B & D
44 options remaining are B & C	20 options remaining are A & C
45 options remaining are A & B	21 options remaining are B & C
46 options remaining are A & D	22 options remaining are A & D
47 options remaining are C & D	23 options remaining are B & C
48 options remaining are B & D	24 options remaining are A & B
49 options remaining are B & C	25 options remaining are B & C
50 options remaining are B & D	26 options remaining are A & D
	27 options remaining are A & C

£250,000

	28 options remaining are A & B
1 options remaining are B & D	29 options remaining are B & C
2 options remaining are B & C	30 options remaining are B & D
3 options remaining are B & C	31 options remaining are A & B
4 options remaining are A & D	32 options remaining are A & C
5 options remaining are B & C	33 options remaining are B & C
6 options remaining are B & D	34 options remaining are A & B
7 options remaining are C & D	35 options remaining are B & C
8 options remaining are A & C	36 options remaining are A & D
9 options remaining are A & B	37 options remaining are A & B
10 options remaining are B & C	38 options remaining are B & C
11 options remaining are B & C	39 options remaining are C & D
12 options remaining are A & C	40 options remaining are A & C
13 options remaining are B & D	41 options remaining are A & D

50:50

£250,000

42 options remaining are A & C
43 options remaining are A & C
44 options remaining are A & C
45 options remaining are A & B
46 options remaining are A & D
47 options remaining are A & C
48 options remaining are B & D
49 options remaining are A & B
50 options remaining are A & C

£500,000

1 options remaining are A & C
2 options remaining are B & C
3 options remaining are A & D
4 options remaining are B & C
5 options remaining are C & D
6 options remaining are A & D
7 options remaining are C & D
8 options remaining are A & D
9 options remaining are A & C
10 options remaining are A & B
11 options remaining are B & C
12 options remaining are B & D
13 options remaining are A & B
14 options remaining are A & D
15 options remaining are B & C
16 options remaining are C & D
17 options remaining are A & B
18 options remaining are A & D
19 options remaining are B & D
20 options remaining are B & C
21 options remaining are A & C
22 options remaining are A & D
23 options remaining are A & B
24 options remaining are C & D
25 options remaining are A & C
26 options remaining are A & B
27 options remaining are A & C
28 options remaining are A & B
29 options remaining are B & C
30 options remaining are B & D
31 options remaining are A & D
32 options remaining are B & C
33 options remaining are A & B
34 options remaining are A & C
35 options remaining are C & D
36 options remaining are A & C
37 options remaining are B & D
38 options remaining are A & C
39 options remaining are B & C
40 options remaining are C & D
41 options remaining are A & B
42 options remaining are B & C
43 options remaining are C & D
44 options remaining are A & C
45 options remaining are C & D

50:50

£500,000

46 options remaining are A & C
47 options remaining are A & B
48 options remaining are B & D
49 options remaining are B & D
50 options remaining are A & C

£1,000,000

1 options remaining are A & B
2 options remaining are C & D
3 options remaining are A & C
4 options remaining are C & D
5 options remaining are A & C
6 options remaining are B & D
7 options remaining are B & C
8 options remaining are A & D
9 options remaining are B & C
10 options remaining are C & D
11 options remaining are A & C
12 options remaining are B & D
13 options remaining are A & C
14 options remaining are B & D
15 options remaining are C & D
16 options remaining are A & B
17 options remaining are C & D
18 options remaining are A & C
19 options remaining are B & C
20 options remaining are A & D
21 options remaining are C & D
22 options remaining are B & D

23 options remaining are A & D
24 options remaining are B & C
25 options remaining are A & D
26 options remaining are B & D
27 options remaining are A & C
28 options remaining are A & D
29 options remaining are B & C
30 options remaining are B & D
31 options remaining are A & C
32 options remaining are C & D
33 options remaining are A & B
34 options remaining are A & D
35 options remaining are B & C
36 options remaining are C & D
37 options remaining are A & D
38 options remaining are B & C
39 options remaining are A & C
40 options remaining are A & B
41 options remaining are B & D
42 options remaining are A & B
43 options remaining are A & D
44 options remaining are B & C
45 options remaining are A & C
46 options remaining are A & D
47 options remaining are A & C
48 options remaining are C & D
49 options remaining are B & D
50 options remaining are C & D

ASK THE AUDIENCE

£100

1 A: 0% B: 1% C: 0% D: 99%

2 A: 3% B: 0% C: 97% D: 0%

3 A: 0% B: 100% C: 0% D: 0%

4 A: 0% B: 0% C: 0% D: 100%

5 A: 0% B: 0% C: 99% D: 1%

6 A: 0% B: 15% C: 4% D: 81%

7 A: 0% B: 0% C: 100% D: 0%

8 A: 100% B: 0% C: 0% D: 0%

9 A: 0% B: 0% C: 2% D: 98%

10 A: 0% B: 100% C: 0% D: 0%

11 A: 3% B: 0% C: 97% D: 0%

12 A: 100% B: 0% C: 0% D: 0%

13 A: 1% B: 99% C: 0% D: 0%

14 A: 0% B: 0% C: 0% D: 100%

15 A: 0% B: 0% C: 100% D: 0%

16 A: 99% B: 1% C: 0% D: 0%

17 A: 0% B: 100% C: 0% D: 0%

18 A: 0% B: 98% C: 0% D: 2%

19 A: 0% B: 0% C: 99% D: 1%

20 A: 100% B: 0% C: 0% D: 0%

21 A: 0% B: 3% C: 0% D: 97%

22 A: 100% B: 0% C: 0% D: 0%

23 A: 0% B: 0% C: 100% D: 0%

24 A: 0% B: 100% C: 0% D: 0%

25 A: 100% B: 0% C: 0% D: 0%

26 A: 100% B: 0% C: 0% D: 0%

27 A: 100% B: 0% C: 0% D: 0%

28 A: 0% B: 100% C: 0% D: 0%

29 A: 2% B: 98% C: 0% D: 0%

30 A: 100% B: 0% C: 0% D: 0%

31 A: 0% B: 3% C: 97% D: 0%

32 A: 0% B: 100% C: 0% D: 0%

33 A: 0% B: 0% C: 0% D: 100%

34 A: 9% B: 0% C: 66% D: 25%

35 A: 0% B: 100% C: 0% D: 0%

36 A: 100% B: 0% C: 0% D: 0%

37 A: 100% B: 0% C: 0% D: 0%

38 A: 0% B: 100% C: 0% D: 0%

39 A: 98% B: 1% C: 0% D: 1%

40 A: 100% B: 0% C: 0% D: 0%

41 A: 0% B: 0% C: 100% D: 0%

42 A: 0% B: 0% C: 100% D: 0%

43 A: 0% B: 100% C: 0% D: 0%

44 A: 96% B: 0% C: 0% D: 4%

45 A: 0% B: 3% C: 97% D: 0%

46 A: 0% B: 100% C: 0% D: 0%

47 A: 99% B: 1% C: 0% D: 0%

48 A: 0% B: 100% C: 0% D: 0%

49 A: 0% B: 100% C: 0% D: 0%

50 A: 99% B: 1% C: 0% D: 0%

51 A: 0% B: 0% C: 100% D: 0%

52 A: 0% B: 100% C: 0% D: 0%

53 A: 2% B: 97% C: 1% D: 0%

54 A: 100% B: 0% C: 0% D: 0%

55 A: 0% B: 0% C: 100% D: 0%

56 A: 0% B: 0% C: 0% D: 100%

57 A: 1% B: 0% C: 0% D: 99%

58 A: 0% B: 100% C: 0% D: 0%

ASK THE AUDIENCE

£100

59 A: 100% B: 0% C: 0% D: 0%

60 A: 99% B: 0% C: 1% D: 0%

61 A: 100% B: 0% C: 0% D: 0%

62 A: 0% B: 100% C: 0% D: 0%

63 A: 0% B: 100% C: 0% D: 0%

64 A: 0% B: 0% C: 100% D: 0%

65 A: 0% B: 100% C: 0% D: 0%

66 A: 0% B: 97% C: 0% D: 3%

67 A: 0% B: 99% C: 1% D: 0%

68 A: 0% B: 0% C: 100% D: 0%

69 A: 0% B: 0% C: 100% D: 0%

70 A: 0% B: 0% C: 100% D: 0%

71 A: 100% B: 0% C: 0% D: 0%

72 A: 99% B: 0% C: 1% D: 0%

73 A: 0% B: 0% C: 100% D: 0%

74 A: 0% B: 0% C: 100% D: 0%

75 A: 0% B: 0% C: 0% D: 100%

76 A: 0% B: 0% C: 100% D: 0%

77 A: 0% B: 0% C: 0% D: 100%

78 A: 0% B: 0% C: 0% D: 100%

79 A: 100% B: 0% C: 0% D: 0%

80 A: 1% B: 99% C: 0% D: 0%

81 A: 3% B: 0% C: 97% D: 0%

82 A: 0% B: 100% C: 0% D: 0%

83 A: 0% B: 0% C: 0% D: 100%

84 A: 0% B: 0% C: 0% D: 100%

85 A: 0% B: 0% C: 100% D: 0%

86 A: 1% B: 99% C: 0% D: 0%

87 A: 0% B: 0% C: 100% D: 0%

88 A: 97% B: 0% C: 0% D: 3%

89 A: 0% B: 100% C: 0% D: 0%

90 A: 99% B: 1% C: 0% D: 0%

91 A: 0% B: 2% C: 98% D: 0%

92 A: 0% B: 0% C: 99% D: 1%

93 A: 0% B: 100% C: 0% D: 0%

94 A: 100% B: 0% C: 0% D: 0%

95 A: 99% B: 0% C: 1% D: 0%

96 A: 0% B: 0% C: 100% D: 0%

97 A: 2% B: 0% C: 98% D: 0%

98 A: 100% B: 0% C: 0% D: 0%

99 A: 0% B: 100% C: 0% D: 0%

100 A: 100% B: 0% C: 0% D: 0%

£200

1 A: 100% B: 0% C: 0% D: 0%

2 A: 0% B: 0% C: 100% D: 0%

3 A: 0% B: 0% C: 100% D: 0%

4 A: 1% B: 0% C: 0% D: 99%

5 A: 0% B: 2% C: 98% D: 0%

6 A: 96% B: 0% C: 4% D: 0%

7 A: 0% B: 100% C: 0% D: 0%

8 A: 0% B: 0% C: 100% D: 0%

9 A: 0% B: 0% C: 100% D: 0%

10 A: 0% B: 0% C: 100% D: 0%

11 A: 1% B: 2% C: 97% D: 0%

12 A: 0% B: 0% C: 0% D: 100%

13 A: 0% B: 2% C: 0% D: 98%

14 A: 4% B: 0% C: 96% D: 0%

£200

15 A: 0% B: 95% C: 0% D: 5%

16 A: 100% B: 0% C: 0% D: 0%

17 A: 0% B: 0% C: 0% D: 100%

18 A: 2% B: 0% C: 0% D: 98%

19 A: 99% B: 0% C: 1% D: 0%

20 A: 0% B: 0% C: 0% D: 100%

21 A: 99% B: 1% C: 0% D: 0%

22 A: 2% B: 0% C: 0% D: 98%

23 A: 96% B: 0% C: 4% D: 0%

24 A: 0% B: 0% C: 100% D: 0%

25 A: 99% B: 0% C: 1% D: 0%

26 A: 97% B: 3% C: 0% D: 0%

27 A: 0% B: 0% C: 100% D: 0%

28 A: 100% B: 0% C: 0% D: 0%

29 A: 0% B: 100% C: 0% D: 0%

30 A: 1% B: 0% C: 0% D: 99%

31 A: 0% B: 2% C: 0% D: 98%

32 A: 0% B: 100% C: 0% D: 0%

33 A: 99% B: 0% C: 1% D: 0%

34 A: 0% B: 100% C: 0% D: 0%

35 A: 0% B: 99% C: 0% D: 1%

36 A: 100% B: 0% C: 0% D: 0%

37 A: 1% B: 98% C: 0% D: 1%

38 A: 1% B: 0% C: 0% D: 99%

39 A: 0% B: 4% C: 0% D: 96%

40 A: 0% B: 100% C: 0% D: 0%

41 A: 0% B: 3% C: 95% D: 2%

42 A: 0% B: 98% C: 2% D: 0%

43 A: 99% B: 0% C: 0% D: 1%

44 A: 11% B: 0% C: 89% D: 0%

45 A: 0% B: 1% C: 99% D: 0%

46 A: 0% B: 86% C: 14% D: 0%

47 A: 100% B: 0% C: 0% D: 0%

48 A: 100% B: 0% C: 0% D: 0%

49 A: 0% B: 1% C: 99% D: 0%

50 A: 98% B: 0% C: 0% D: 2%

51 A: 100% B: 0% C: 0% D: 0%

52 A: 5% B: 0% C: 95% D: 0%

53 A: 0% B: 0% C: 0% D: 100%

54 A: 0% B: 0% C: 5% D: 95%

55 A: 2% B: 0% C: 98% D: 0%

56 A: 0% B: 99% C: 1% D: 0%

57 A: 97% B: 0% C: 3% D: 0%

58 A: 2% B: 0% C: 98% D: 0%

59 A: 0% B: 0% C: 0% D: 100%

60 A: 99% B: 0% C: 1% D: 0%

61 A: 96% B: 0% C: 4% D: 0%

62 A: 0% B: 0% C: 100% D: 0%

63 A: 2% B: 98% C: 0% D: 0%

64 A: 0% B: 0% C: 1% D: 99%

65 A: 100% B: 0% C: 0% D: 0%

66 A: 0% B: 98% C: 0% D: 2%

67 A: 96% B: 0% C: 4% D: 0%

68 A: 0% B: 100% C: 0% D: 0%

69 A: 0% B: 0% C: 100% D: 0%

70 A: 0% B: 98% C: 2% D: 0%

71 A: 0% B: 0% C: 100% D: 0%

72 A: 14% B: 86% C: 0% D: 0%

ASK THE AUDIENCE

£200

73 A: 0% B: 100% C: 0% D: 0%
74 A: 98% B: 0% C: 2% D: 0%
75 A: 0% B: 97% C: 0% D: 3%
76 A: 0% B: 1% C: 0% D: 99%
77 A: 100% B: 0% C: 0% D: 0%
78 A: 0% B: 0% C: 100% D: 0%
79 A: 0% B: 1% C: 99% D: 0%
80 A: 0% B: 7% C: 0% D: 93%
81 A: 96% B: 0% C: 0% D: 4%
82 A: 99% B: 0% C: 1% D: 0%
83 A: 95% B: 0% C: 5% D: 0%
84 A: 0% B: 0% C: 100% D: 0%
85 A: 0% B: 96% C: 4% D: 0%
86 A: 0% B: 97% C: 3% D: 0%
87 A: 94% B: 0% C: 0% D: 6%
88 A: 0% B: 0% C: 100% D: 0%
89 A: 0% B: 2% C: 98% D: 0%
90 A: 96% B: 0% C: 4% D: 0%
91 A: 94% B: 6% C: 0% D: 0%
92 A: 1% B: 0% C: 0% D: 99%
93 A: 0% B: 11% C: 89% D: 0%
94 A: 1% B: 0% C: 0% D: 99%
95 A: 0% B: 96% C: 4% D: 0%
96 A: 98% B: 0% C: 0% D: 2%
97 A: 100% B: 0% C: 0% D: 0%
98 A: 0% B: 97% C: 3% D: 0%
99 A: 0% B: 100% C: 0% D: 0%
100 A: 0% B: 97% C: 3% D: 0%

£300

1 A: 0% B: 100% C: 0% D: 0%
2 A: 15% B: 85% C: 0% D: 0%
3 A: 0% B: 100% C: 0% D: 0%
4 A: 0% B: 11% C: 2% D: 87%
5 A: 0% B: 100% C: 0% D: 0%
6 A: 0% B: 0% C: 100% D: 0%
7 A: 0% B: 1% C: 99% D: 0%
8 A: 0% B: 0% C: 0% D: 100%
9 A: 8% B: 0% C: 88% D: 4%
10 A: 0% B: 100% C: 0% D: 0%
11 A: 0% B: 100% C: 0% D: 0%
12 A: 0% B: 7% C: 93% D: 0%
13 A: 1% B: 13% C: 0% D: 86%
14 A: 0% B: 98% C: 0% D: 2%
15 A: 0% B: 0% C: 100% D: 0%
16 A: 1% B: 99% C: 0% D: 0%
17 A: 0% B: 0% C: 0% D: 100%
18 A: 0% B: 99% C: 1% D: 0%
19 A: 0% B: 0% C: 6% D: 94%
20 A: 100% B: 0% C: 0% D: 0%
21 A: 3% B: 19% C: 0% D: 78%
22 A: 100% B: 0% C: 0% D: 0%
23 A: 5% B: 6% C: 0% D: 89%
24 A: 0% B: 100% C: 0% D: 0%
25 A: 98% B: 0% C: 0% D: 2%
26 A: 0% B: 100% C: 0% D: 0%
27 A: 88% B: 9% C: 0% D: 3%
28 A: 0% B: 83% C: 11% D: 6%

ASK THE AUDIENCE

£300

29	A: 0% B: 7% C: 93% D: 0%	57	A: 4% B: 68% C: 28% D: 0%
30	A: 0% B: 0% C: 100% D: 0%	58	A: 44% B: 56% C: 0% D: 0%
31	A: 11% B: 0% C: 0% D: 89%	59	A: 100% B: 0% C: 0% D: 0%
32	A: 0% B: 0% C: 100% D: 0%	60	A: 0% B: 0% C: 100% D: 0%
33	A: 7% B: 27% C: 5% D: 61%	61	A: 0% B: 94% C: 0% D: 6%
34	A: 0% B: 0% C: 0% D: 100%	62	A: 0% B: 100% C: 0% D: 0%
35	A: 0% B: 0% C: 0% D: 100%	63	A: 16% B: 0% C: 84% D: 0%
36	A: 0% B: 0% C: 97% D: 3%	64	A: 100% B: 0% C: 0% D: 0%
37	A: 0% B: 0% C: 100% D: 0%	65	A: 0% B: 92% C: 0% D: 8%
38	A: 0% B: 0% C: 100% D: 0%	66	A: 0% B: 100% C: 0% D: 0%
39	A: 0% B: 5% C: 2% D: 93%	67	A: 0% B: 0% C: 9% D: 91%
40	A: 100% B: 0% C: 0% D: 0%	68	A: 0% B: 0% C: 98% D: 2%
41	A: 0% B: 0% C: 0% D: 100%	69	A: 22% B: 0% C: 78% D: 0%
42	A: 0% B: 83% C: 0% D: 17%	70	A: 0% B: 100% C: 0% D: 0%
43	A: 100% B: 0% C: 0% D: 0%	71	A: 0% B: 33% C: 61% D: 6%
44	A: 0% B: 100% C: 0% D: 0%	72	A: 4% B: 1% C: 95% D: 0%
45	A: 3% B: 0% C: 0% D: 97%	73	A: 0% B: 100% C: 0% D: 0%
46	A: 0% B: 100% C: 0% D: 0%	74	A: 4% B: 0% C: 0% D: 96%
47	A: 17% B: 67% C: 5% D: 11%	75	A: 10% B: 0% C: 0% D: 90%
48	A: 8% B: 6% C: 86% D: 0%	76	A: 100% B: 0% C: 0% D: 0%
49	A: 0% B: 100% C: 0% D: 0%	77	A: 0% B: 67% C: 33% D: 0%
50	A: 0% B: 2% C: 97% D: 1%	78	A: 100% B: 0% C: 0% D: 0%
51	A: 0% B: 0% C: 6% D: 94%	79	A: 85% B: 15% C: 0% D: 0%
52	A: 4% B: 61% C: 8% D: 27%	80	A: 6% B: 0% C: 0% D: 94%
53	A: 0% B: 0% C: 5% D: 95%	81	A: 0% B: 0% C: 100% D: 0%
54	A: 0% B: 10% C: 90% D: 0%	82	A: 2% B: 0% C: 87% D: 11%
55	A: 0% B: 98% C: 2% D: 0%	83	A: 0% B: 0% C: 98% D: 2%
56	A: 100% B: 0% C: 0% D: 0%	84	A: 28% B: 0% C: 4% D: 68%

ASK THE AUDIENCE

£300

85 A: 89% B: 0% C: 0% D: 11%

86 A: 98% B: 0% C: 0% D: 2%

87 A: 0% B: 97% C: 0% D: 3%

88 A: 0% B: 0% C: 100% D: 0%

89 A: 0% B: 0% C: 5% D: 95%

90 A: 0% B: 0% C: 84% D: 16%

91 A: 0% B: 100% C: 0% D: 0%

92 A: 100% B: 0% C: 0% D: 0%

93 A: 6% B: 0% C: 94% D: 0%

94 A: 0% B: 0% C: 99% D: 1%

95 A: 0% B: 0% C: 100% D: 0%

96 A: 0% B: 92% C: 0% D: 8%

97 A: 0% B: 100% C: 0% D: 0%

98 A: 0% B: 96% C: 2% D: 2%

99 A: 0% B: 0% C: 0% D: 100%

100 A: 0% B: 0% C: 99% D: 0%

£500

1 A: 0% B: 0% C: 100% D: 0%

2 A: 98% B: 0% C: 2% D: 0%

3 A: 17% B: 0% C: 0% D: 83%

4 A: 0% B: 0% C: 0% D: 100%

5 A: 5% B: 78% C: 11% D: 6%

6 A: 0% B: 0% C: 0% D: 100%

7 A: 0% B: 0% C: 97% D: 3%

8 A: 100% B: 0% C: 0% D: 0%

9 A: 0% B: 0% C: 100% D: 0%

10 A: 0% B: 99% C: 0% D: 1%

11 A: 0% B: 100% C: 0% D: 0%

12 A: 100% B: 0% C: 0% D: 0%

13 A: 0% B: 17% C: 0% D: 83%

14 A: 0% B: 0% C: 0% D: 100%

15 A: 9% B: 0% C: 91% D: 0%

16 A: 0% B: 100% C: 0% D: 0%

17 A: 0% B: 0% C: 100% D: 0%

18 A: 0% B: 0% C: 0% D: 100%

19 A: 0% B: 0% C: 100% D: 0%

20 A: 0% B: 3% C: 97% D: 0%

21 A: 0% B: 0% C: 100% D: 0%

22 A: 22% B: 0% C: 0% D: 78%

23 A: 22% B: 45% C: 11% D: 22%

24 A: 1% B: 0% C: 0% D: 99%

25 A: 0% B: 0% C: 83% D: 17%

26 A: 0% B: 98% C: 2% D: 0%

27 A: 33% B: 45% C: 0% D: 22%

28 A: 17% B: 22% C: 61% D: 0%

29 A: 0% B: 0% C: 0% D: 100%

30 A: 7% B: 0% C: 5% D: 88%

31 A: 82% B: 7% C: 0% D: 11%

32 A: 100% B: 0% C: 0% D: 0%

33 A: 0% B: 89% C: 0% D: 11%

34 A: 0% B: 99% C: 1% D: 0%

35 A: 0% B: 0% C: 100% D: 0%

36 A: 0% B: 0% C: 100% D: 0%

37 A: 17% B: 0% C: 11% D: 72%

38 A: 0% B: 100% C: 0% D: 0%

39 A: 11% B: 17% C: 61% D: 11%

40 A: 0% B: 4% C: 96% D: 0%

ASK THE AUDIENCE

£500

41 A: 25% B: 3% C: 72% D: 0%

42 A: 0% B: 100% C: 0% D: 0%

43 A: 67% B: 33% C: 0% D: 0%

44 A: 0% B: 0% C: 100% D: 0%

45 A: 13% B: 0% C: 9% D: 78%

46 A: 0% B: 0% C: 100% D: 0%

47 A: 0% B: 100% C: 0% D: 0%

48 A: 94% B: 6% C: 0% D: 0%

49 A: 100% B: 0% C: 0% D: 0%

50 A: 2% B: 0% C: 98% D: 0%

51 A: 0% B: 0% C: 100% D: 0%

52 A: 0% B: 100% C: 0% D: 0%

53 A: 5% B: 16% C: 18% D: 61%

54 A: 0% B: 100% C: 0% D: 0%

55 A: 0% B: 100% C: 0% D: 0%

56 A: 28% B: 5% C: 67% D: 0%

57 A: 0% B: 0% C: 0% D: 100%

58 A: 0% B: 0% C: 100% D: 0%

59 A: 67% B: 13% C: 0% D: 20%

60 A: 0% B: 0% C: 0% D: 100%

61 A: 22% B: 0% C: 3% D: 75%

62 A: 0% B: 0% C: 100% D: 0%

63 A: 0% B: 0% C: 15% D: 85%

64 A: 0% B: 0% C: 100% D: 0%

65 A: 100% B: 0% C: 0% D: 0%

66 A: 0% B: 8% C: 92% D: 0%

67 A: 0% B: 0% C: 0% D: 100%

68 A: 19% B: 73% C: 0% D: 8%

69 A: 0% B: 0% C: 0% D: 100%

70 A: 0% B: 0% C: 100% D: 0%

71 A: 0% B: 0% C: 100% D: 0%

72 A: 0% B: 78% C: 0% D: 22%

73 A: 100% B: 0% C: 0% D: 0%

74 A: 3% B: 0% C: 0% D: 97%

75 A: 0% B: 0% C: 1% D: 99%

76 A: 100% B: 0% C: 0% D: 0%

77 A: 5% B: 0% C: 12% D: 83%

78 A: 0% B: 100% C: 0% D: 0%

79 A: 0% B: 0% C: 0% D: 100%

80 A: 0% B: 99% C: 1% D: 0%

81 A: 0% B: 100% C: 0% D: 0%

82 A: 100% B: 0% C: 0% D: 0%

83 A: 71% B: 5% C: 7% D: 17%

84 A: 0% B: 100% C: 0% D: 0%

85 A: 0% B: 0% C: 0% D: 100%

86 A: 0% B: 0% C: 100% D: 0%

87 A: 0% B: 97% C: 3% D: 0%

88 A: 100% B: 0% C: 0% D: 0%

89 A: 0% B: 100% C: 0% D: 0%

90 A: 0% B: 10% C: 6% D: 84%

91 A: 0% B: 0% C: 100% D: 0%

92 A: 0% B: 100% C: 0% D: 0%

93 A: 0% B: 97% C: 0% D: 3%

94 A: 2% B: 98% C: 0% D: 0%

95 A: 0% B: 0% C: 100% D: 0%

96 A: 16% B: 79% C: 0% D: 5%

97 A: 0% B: 0% C: 0% D: 100%

98 A: 0% B: 0% C: 3% D: 97%

ASK THE AUDIENCE

£500

99 A: 0% B: 0% C: 0% D: 100%

100 A: 3% B: 0% C: 0% D: 97%

£1,000

1 A: 0% B: 100% C: 0% D: 0%

2 A: 25% B: 3% C: 0% D: 72%

3 A: 0% B: 1% C: 0% D: 99%

4 A: 0% B: 99% C: 0% D: 1%

5 A: 7% B: 8% C: 82% D: 3%

6 A: 0% B: 100% C: 0% D: 0%

7 A: 0% B: 5% C: 95% D: 0%

8 A: 5% B: 31% C: 0% D: 64%

9 A: 74% B: 6% C: 0% D: 20%

10 A: 0% B: 99% C: 1% D: 0%

11 A: 0% B: 0% C: 21% D: 79%

12 A: 71% B: 8% C: 0% D: 21%

13 A: 0% B: 0% C: 0% D: 100%

14 A: 21% B: 0% C: 0% D: 79%

15 A: 0% B: 100% C: 0% D: 0%

16 A: 86% B: 8% C: 0% D: 6%

17 A: 1% B: 0% C: 98% D: 1%

18 A: 93% B: 7% C: 0% D: 0%

19 A: 4% B: 0% C: 96% D: 0%

20 A: 0% B: 1% C: 99% D: 0%

21 A: 9% B: 77% C: 3% D: 11%

22 A: 12% B: 4% C: 84% D: 0%

23 A: 0% B: 0% C: 97% D: 3%

24 A: 0% B: 0% C: 100% D: 0%

25 A: 3% B: 0% C: 14% D: 83%

26 A: 2% B: 0% C: 0% D: 98%

27 A: 31% B: 0% C: 57% D: 12%

28 A: 16% B: 6% C: 72% D: 6%

29 A: 89% B: 11% C: 0% D: 0%

30 A: 8% B: 42% C: 42% D: 8%

31 A: 15% B: 76% C: 0% D: 9%

32 A: 1% B: 99% C: 0% D: 0%

33 A: 0% B: 99% C: 0% D: 1%

34 A: 26% B: 56% C: 2% D: 16%

35 A: 10% B: 34% C: 0% D: 56%

36 A: 0% B: 71% C: 0% D: 29%

37 A: 0% B: 3% C: 97% D: 0%

38 A: 0% B: 52% C: 48% D: 0%

39 A: 11% B: 0% C: 72% D: 17%

40 A: 36% B: 43% C: 0% D: 21%

41 A: 8% B: 9% C: 83% D: 0%

42 A: 98% B: 2% C: 0% D: 0%

43 A: 93% B: 0% C: 0% D: 7%

44 A: 84% B: 16% C: 0% D: 0%

45 A: 9% B: 73% C: 13% D: 5%

46 A: 36% B: 21% C: 43% D: 0%

47 A: 0% B: 91% C: 0% D: 9%

48 A: 79% B: 0% C: 0% D: 21%

49 A: 4% B: 10% C: 0% D: 86%

50 A: 4% B: 96% C: 0% D: 0%

51 A: 0% B: 21% C: 25% D: 54%

52 A: 36% B: 4% C: 43% D: 17%

53 A: 8% B: 54% C: 29% D: 9%

54 A: 0% B: 96% C: 0% D: 4%

ASK THE AUDIENCE

£1,000

55 A: 0% B: 0% C: 92% D: 8%
56 A: 11% B: 15% C: 11% D: 63%
57 A: 2% B: 97% C: 0% D: 1%
58 A: 93% B: 4% C: 3% D: 0%
59 A: 6% B: 13% C: 0% D: 81%
60 A: 3% B: 54% C: 0% D: 43%
61 A: 9% B: 0% C: 83% D: 8%
62 A: 0% B: 2% C: 0% D: 98%
63 A: 0% B: 11% C: 72% D: 17%
64 A: 0% B: 99% C: 1% D: 0%
65 A: 9% B: 12% C: 7% D: 72%
66 A: 79% B: 21% C: 0% D: 0%
67 A: 3% B: 0% C: 54% D: 43%
68 A: 8% B: 7% C: 79% D: 6%
69 A: 91% B: 0% C: 9% D: 0%
70 A: 12% B: 0% C: 79% D: 9%
71 A: 16% B: 0% C: 72% D: 12%
72 A: 29% B: 71% C: 0% D: 0%
73 A: 9% B: 86% C: 5% D: 0%
74 A: 0% B: 0% C: 8% D: 92%
75 A: 4% B: 0% C: 85% D: 11%
76 A: 83% B: 0% C: 5% D: 12%
77 A: 0% B: 71% C: 0% D: 29%
78 A: 10% B: 0% C: 55% D: 35%
79 A: 2% B: 93% C: 5% D: 0%
80 A: 3% B: 0% C: 0% D: 97%
81 A: 13% B: 9% C: 5% D: 73%
82 A: 96% B: 4% C: 0% D: 0%
83 A: 0% B: 54% C: 25% D: 21%

84 A: 15% B: 0% C: 9% D: 76%
85 A: 0% B: 96% C: 4% D: 0%
86 A: 9% B: 0% C: 91% D: 0%
87 A: 0% B: 96% C: 4% D: 0%
88 A: 2% B: 6% C: 92% D: 0%
89 A: 97% B: 2% C: 1% D: 0%
90 A: 1% B: 3% C: 95% D: 1%
91 A: 0% B: 11% C: 89% D: 0%
92 A: 0% B: 88% C: 0% D: 12%
93 A: 89% B: 11% C: 0% D: 0%
94 A: 0% B: 2% C: 93% D: 5%
95 A: 8% B: 12% C: 71% D: 9%
96 A: 0% B: 0% C: 99% D: 1%
97 A: 1% B: 84% C: 6% D: 9%
98 A: 93% B: 7% C: 0% D: 0%
99 A: 0% B: 8% C: 84% D: 8%
100 A: 90% B: 5% C: 5% D: 0%

£2,000

1 A: 0% B: 97% C: 3% D: 0%
2 A: 5% B: 68% C: 23% D: 4%
3 A: 0% B: 3% C: 0% D: 97%
4 A: 91% B: 9% C: 0% D: 0%
5 A: 1% B: 5% C: 76% D: 18%
6 A: 0% B: 0% C: 95% D: 5%
7 A: 88% B: 5% C: 7% D: 0%
8 A: 87% B: 0% C: 0% D: 13%
9 A: 71% B: 15% C: 0% D: 14%
10 A: 82% B: 13% C: 5% D: 0%

ASK THE AUDIENCE

£2,000

11 A: 9% B: 0% C: 86% D: 5%
12 A: 9% B: 5% C: 0% D: 86%
13 A: 92% B: 0% C: 5% D: 3%
14 A: 3% B: 0% C: 82% D: 15%
15 A: 5% B: 0% C: 77% D: 18%
16 A: 89% B: 5% C: 0% D: 6%
17 A: 18% B: 9% C: 0% D: 73%
18 A: 1% B: 0% C: 99% D: 0%
19 A: 85% B: 0% C: 12% D: 3%
20 A: 97% B: 0% C: 0% D: 3%
21 A: 6% B: 21% C: 64% D: 9%
22 A: 5% B: 73% C: 4% D: 18%
23 A: 82% B: 0% C: 15% D: 3%
24 A: 5% B: 72% C: 4% D: 19%
25 A: 0% B: 0% C: 98% D: 2%
26 A: 4% B: 81% C: 5% D: 10%
27 A: 95% B: 0% C: 5% D: 0%
28 A: 0% B: 0% C: 9% D: 91%
29 A: 5% B: 0% C: 91% D: 4%
30 A: 23% B: 17% C: 6% D: 54%
31 A: 0% B: 91% C: 5% D: 4%
32 A: 8% B: 92% C: 0% D: 0%
33 A: 4% B: 0% C: 4% D: 92%
34 A: 3% B: 16% C: 3% D: 78%
35 A: 91% B: 2% C: 0% D: 7%
36 A: 77% B: 5% C: 14% D: 4%
37 A: 18% B: 0% C: 77% D: 5%
38 A: 0% B: 9% C: 6% D: 85%
39 A: 7% B: 8% C: 77% D: 8%
40 A: 7% B: 0% C: 0% D: 93%
41 A: 94% B: 0% C: 0% D: 6%
42 A: 5% B: 27% C: 59% D: 9%
43 A: 98% B: 0% C: 1% D: 1%
44 A: 9% B: 5% C: 0% D: 86%
45 A: 0% B: 89% C: 0% D: 11%
46 A: 99% B: 0% C: 1% D: 0%
47 A: 99% B: 0% C: 1% D: 0%
48 A: 4% B: 75% C: 21% D: 0%
49 A: 89% B: 4% C: 0% D: 7%
50 A: 89% B: 3% C: 1% D: 7%
51 A: 3% B: 92% C: 2% D: 3%
52 A: 6% B: 94% C: 0% D: 0%
53 A: 7% B: 0% C: 88% D: 5%
54 A: 0% B: 0% C: 100% D: 0%
55 A: 95% B: 0% C: 0% D: 5%
56 A: 77% B: 4% C: 3% D: 16%
57 A: 4% B: 91% C: 5% D: 0%
58 A: 15% B: 83% C: 0% D: 2%
59 A: 1% B: 97% C: 0% D: 2%
60 A: 4% B: 0% C: 96% D: 0%
61 A: 4% B: 0% C: 5% D: 91%
62 A: 2% B: 0% C: 0% D: 98%
63 A: 4% B: 0% C: 91% D: 5%
64 A: 0% B: 91% C: 0% D: 9%
65 A: 8% B: 0% C: 92% D: 0%
66 A: 10% B: 64% C: 16% D: 10%
67 A: 9% B: 32% C: 0% D: 59%
68 A: 84% B: 9% C: 0% D: 7%

ASK THE AUDIENCE

£2,000

69 A: 8% B: 7% C: 82% D: 3%
70 A: 3% B: 6% C: 91% D: 0%
71 A: 7% B: 3% C: 15% D: 75%
72 A: 91% B: 6% C: 0% D: 3%
73 A: 2% B: 8% C: 63% D: 27%
74 A: 5% B: 0% C: 36% D: 59%
75 A: 41% B: 18% C: 5% D: 36%

£4,000

 1 A: 13% B: 87% C: 0% D: 0%
 2 A: 15% B: 54% C: 0% D: 31%
 3 A: 80% B: 7% C: 13% D: 0%
 4 A: 9% B: 0% C: 91% D: 0%
 5 A: 2% B: 0% C: 98% D: 0%
 6 A: 10% B: 0% C: 79% D: 11%
 7 A: 3% B: 0% C: 97% D: 0%
 8 A: 4% B: 92% C: 4% D: 0%
 9 A: 10% B: 4% C: 0% D: 86%
10 A: 79% B: 11% C: 0% D: 10%
11 A: 7% B: 0% C: 9% D: 84%
12 A: 18% B: 0% C: 75% D: 7%
13 A: 11% B: 87% C: 0% D: 2%
14 A: 96% B: 1% C: 0% D: 3%
15 A: 13% B: 60% C: 27% D: 0%
16 A: 1% B: 0% C: 99% D: 0%
17 A: 0% B: 9% C: 91% D: 0%
18 A: 7% B: 0% C: 93% D: 0%
19 A: 88% B: 7% C: 0% D: 5%
20 A: 22% B: 75% C: 0% D: 3%

21 A: 17% B: 39% C: 41% D: 3%
22 A: 3% B: 87% C: 0% D: 10%
23 A: 7% B: 44% C: 46% D: 3%
24 A: 12% B: 0% C: 0% D: 88%
25 A: 2% B: 0% C: 98% D: 0%
26 A: 15% B: 52% C: 18% D: 15%
27 A: 1% B: 1% C: 98% D: 0%
28 A: 1% B: 0% C: 0% D: 99%
29 A: 5% B: 7% C: 1% D: 87%
30 A: 93% B: 7% C: 0% D: 0%
31 A: 0% B: 0% C: 17% D: 83%
32 A: 76% B: 19% C: 0% D: 5%
33 A: 85% B: 3% C: 12% D: 0%
34 A: 4% B: 15% C: 0% D: 81%
35 A: 6% B: 88% C: 0% D: 6%
36 A: 6% B: 13% C: 74% D: 7%
37 A: 4% B: 79% C: 11% D: 6%
38 A: 13% B: 23% C: 4% D: 60%
39 A: 2% B: 29% C: 3% D: 66%
40 A: 28% B: 0% C: 67% D: 5%
41 A: 1% B: 87% C: 7% D: 5%
42 A: 16% B: 0% C: 84% D: 0%
43 A: 0% B: 0% C: 94% D: 6%
44 A: 9% B: 87% C: 0% D: 4%
45 A: 6% B: 5% C: 0% D: 89%
46 A: 45% B: 0% C: 8% D: 47%
47 A: 33% B: 64% C: 0% D: 3%
48 A: 98% B: 2% C: 0% D: 0%
49 A: 9% B: 91% C: 0% D: 0%

ASK THE AUDIENCE

£4,000

50	A: 0%	B: 4%	C: 0%	D: 96%
51	A: 20%	B: 1%	C: 76%	D: 3%
52	A: 5%	B: 3%	C: 92%	D: 0%
53	A: 0%	B: 99%	C: 1%	D: 0%
54	A: 15%	B: 18%	C: 52%	D: 15%
55	A: 24%	B: 63%	C: 0%	D: 13%
56	A: 24%	B: 5%	C: 71%	D: 0%
57	A: 7%	B: 83%	C: 0%	D: 10%
58	A: 11%	B: 55%	C: 30%	D: 4%
59	A: 15%	B: 76%	C: 0%	D: 9%
60	A: 1%	B: 0%	C: 99%	D: 0%
61	A: 98%	B: 2%	C: 0%	D: 0%
62	A: 0%	B: 15%	C: 0%	D: 85%
63	A: 3%	B: 0%	C: 4%	D: 93%
64	A: 15%	B: 23%	C: 5%	D: 57%
65	A: 52%	B: 35%	C: 10%	D: 3%
66	A: 0%	B: 9%	C: 91%	D: 0%
67	A: 10%	B: 17%	C: 0%	D: 73%
68	A: 8%	B: 83%	C: 5%	D: 4%
69	A: 0%	B: 0%	C: 1%	D: 99%
70	A: 65%	B: 5%	C: 12%	D: 18%
71	A: 92%	B: 8%	C: 0%	D: 0%
72	A: 87%	B: 5%	C: 4%	D: 4%
73	A: 5%	B: 0%	C: 5%	D: 90%
74	A: 9%	B: 11%	C: 19%	D: 61%
75	A: 8%	B: 0%	C: 92%	D: 0%

£8,000

1	A: 32%	B: 0%	C: 68%	D: 0%
2	A: 79%	B: 8%	C: 0%	D: 13%
3	A: 11%	B: 17%	C: 72%	D: 0%
4	A: 0%	B: 77%	C: 14%	D: 9%
5	A: 0%	B: 93%	C: 0%	D: 7%
6	A: 23%	B: 11%	C: 59%	D: 7%
7	A: 8%	B: 0%	C: 86%	D: 6%
8	A: 15%	B: 63%	C: 13%	D: 9%
9	A: 82%	B: 0%	C: 4%	D: 14%
10	A: 5%	B: 87%	C: 8%	D: 0%
11	A: 14%	B: 82%	C: 0%	D: 4%
12	A: 5%	B: 41%	C: 2%	D: 52%
13	A: 97%	B: 0%	C: 3%	D: 0%
14	A: 64%	B: 17%	C: 10%	D: 9%
15	A: 9%	B: 0%	C: 5%	D: 86%
16	A: 59%	B: 4%	C: 5%	D: 32%
17	A: 3%	B: 7%	C: 0%	D: 90%
18	A: 11%	B: 43%	C: 35%	D: 11%
19	A: 51%	B: 13%	C: 4%	D: 32%
20	A: 19%	B: 68%	C: 0%	D: 13%
21	A: 6%	B: 9%	C: 0%	D: 85%
22	A: 75%	B: 19%	C: 6%	D: 0%
23	A: 9%	B: 79%	C: 4%	D: 8%
24	A: 6%	B: 9%	C: 67%	D: 18%
25	A: 16%	B: 4%	C: 0%	D: 80%
26	A: 18%	B: 72%	C: 10%	D: 0%
27	A: 84%	B: 7%	C: 9%	D: 0%
28	A: 5%	B: 5%	C: 82%	D: 5%
29	A: 73%	B: 4%	C: 18%	D: 5%
30	A: 13%	B: 2%	C: 85%	D: 0%

ASK THE AUDIENCE

£8,000

31 A: 4% B: 88% C: 8% D: 0%

32 A: 91% B: 4% C: 5% D: 0%

33 A: 13% B: 28% C: 18% D: 41%

34 A: 10% B: 9% C: 0% D: 81%

35 A: 91% B: 4% C: 5% D: 0%

36 A: 0% B: 0% C: 93% D: 7%

37 A: 10% B: 0% C: 90% D: 0%

38 A: 32% B: 13% C: 14% D: 41%

39 A: 18% B: 66% C: 5% D: 11%

40 A: 0% B: 6% C: 0% D: 94%

41 A: 0% B: 0% C: 95% D: 5%

42 A: 0% B: 8% C: 0% D: 92%

43 A: 4% B: 14% C: 59% D: 23%

44 A: 80% B: 6% C: 7% D: 7%

45 A: 81% B: 12% C: 7% D: 0%

46 A: 80% B: 6% C: 7% D: 7%

47 A: 9% B: 82% C: 0% D: 9%

48 A: 76% B: 10% C: 14% D: 0%

49 A: 27% B: 73% C: 0% D: 0%

50 A: 3% B: 24% C: 73% D: 0%

51 A: 14% B: 3% C: 4% D: 79%

52 A: 5% B: 0% C: 87% D: 8%

53 A: 12% B: 0% C: 18% D: 70%

54 A: 4% B: 89% C: 0% D: 7%

55 A: 98% B: 2% C: 0% D: 0%

56 A: 0% B: 53% C: 36% D: 11%

57 A: 4% B: 31% C: 65% D: 0%

58 A: 9% B: 9% C: 4% D: 78%

59 A: 0% B: 26% C: 5% D: 69%

60 A: 0% B: 94% C: 6% D: 0%

61 A: 18% B: 0% C: 12% D: 70%

62 A: 0% B: 0% C: 97% D: 3%

63 A: 79% B: 13% C: 0% D: 8%

64 A: 73% B: 0% C: 0% D: 27%

65 A: 9% B: 14% C: 77% D: 0%

66 A: 73% B: 18% C: 4% D: 5%

67 A: 14% B: 76% C: 0% D: 10%

68 A: 82% B: 14% C: 4% D: 0%

69 A: 4% B: 8% C: 0% D: 88%

70 A: 4% B: 14% C: 0% D: 82%

71 A: 0% B: 93% C: 0% D: 7%

72 A: 5% B: 0% C: 90% D: 5%

73 A: 80% B: 5% C: 7% D: 8%

74 A: 91% B: 7% C: 0% D: 0%

75 A: 0% B: 0% C: 88% D: 12%

£16,000

1 A: 32% B: 0% C: 68% D: 0%

2 A: 8% B: 3% C: 0% D: 89%

3 A: 5% B: 0% C: 75% D: 20%

4 A: 8% B: 79% C: 0% D: 13%

5 A: 13% B: 19% C: 59% D: 9%

6 A: 29% B: 34% C: 33% D: 4%

7 A: 0% B: 72% C: 7% D: 21%

8 A: 28% B: 67% C: 0% D: 5%

9 A: 66% B: 9% C: 8% D: 17%

10 A: 4% B: 21% C: 75% D: 0%

11 A: 12% B: 55% C: 13% D: 20%

ASK THE AUDIENCE

£16,000

12 A: 76% B: 11% C: 0% D: 13%

13 A: 12% B: 64% C: 3% D: 12%

14 A: 7% B: 0% C: 84% D: 9%

15 A: 0% B: 67% C: 18% D: 15%

16 A: 12% B: 20% C: 14% D: 54%

17 A: 66% B: 25% C: 9% D: 0%

18 A: 40% B: 17% C: 13% D: 30%

19 A: 59% B: 13% C: 0% D: 28%

20 A: 9% B: 0% C: 79% D: 12%

21 A: 4% B: 63% C: 4% D: 29%

22 A: 89% B: 0% C: 0% D: 11%

23 A: 0% B: 66% C: 8% D: 26%

24 A: 23% B: 68% C: 3% D: 6%

25 A: 25% B: 18% C: 6% D: 51%

26 A: 9% B: 7% C: 5% D: 79%

27 A: 71% B: 15% C: 1% D: 13%

28 A: 16% B: 11% C: 14% D: 59%

29 A: 22% B: 13% C: 58% D: 7%

30 A: 0% B: 0% C: 9% D: 91%

31 A: 6% B: 73% C: 10% D: 11%

32 A: 3% B: 28% C: 5% D: 64%

33 A: 72% B: 1% C: 24% D: 3%

34 A: 42% B: 19% C: 2% D: 37%

35 A: 8% B: 13% C: 10% D: 69%

36 A: 15% B: 6% C: 79% D: 0%

37 A: 4% B: 16% C: 9% D: 71%

38 A: 21% B: 3% C: 12% D: 64%

39 A: 20% B: 7% C: 25% D: 48%

40 A: 9% B: 0% C: 87% D: 4%

41 A: 13% B: 12% C: 8% D: 67%

42 A: 0% B: 15% C: 85% D: 0%

43 A: 6% B: 9% C: 3% D: 82%

44 A: 16% B: 6% C: 74% D: 4%

45 A: 18% B: 0% C: 24% D: 58%

46 A: 0% B: 89% C: 4% D: 7%

47 A: 1% B: 85% C: 11% D: 3%

48 A: 77% B: 6% C: 0% D: 17%

49 A: 7% B: 5% C: 88% D: 0%

50 A: 0% B: 0% C: 91% D: 9%

51 A: 12% B: 13% C: 22% D: 53%

52 A: 13% B: 76% C: 3% D: 8%

53 A: 21% B: 5% C: 25% D: 49%

54 A: 51% B: 33% C: 4% D: 12%

55 A: 0% B: 82% C: 9% D: 9%

56 A: 61% B: 10% C: 29% D: 0%

57 A: 13% B: 12% C: 67% D: 8%

58 A: 10% B: 73% C: 6% D: 11%

59 A: 6% B: 16% C: 4% D: 74%

60 A: 1% B: 85% C: 3% D: 11%

61 A: 76% B: 3% C: 13% D: 8%

62 A: 15% B: 0% C: 67% D: 18%

63 A: 4% B: 5% C: 91% D: 0%

64 A: 0% B: 68% C: 0% D: 32%

65 A: 0% B: 3% C: 8% D: 89%

66 A: 8% B: 0% C: 13% D: 79%

67 A: 11% B: 76% C: 0% D: 13%

68 A: 0% B: 87% C: 9% D: 4%

69 A: 71% B: 1% C: 13% D: 15%

ASK THE AUDIENCE

£16,000

70 A: 79% B: 12% C: 9% D: 0%
71 A: 24% B: 3% C: 1% D: 72%
72 A: 12% B: 51% C: 33% D: 4%
73 A: 64% B: 3% C: 21% D: 12%
74 A: 0% B: 91% C: 9% D: 0%
75 A: 3% B: 0% C: 8% D: 89%

£32,000

1   A: 30% B: 13% C: 14% D: 43%
2   A: 29% B: 21% C: 3% D: 47%
3   A: 36% B: 15% C: 13% D: 36%
4   A: 13% B: 63% C: 24% D: 0%
5   A: 0% B: 34% C: 6% D: 60%
6   A: 0% B: 29% C: 71% D: 0%
7   A: 14% B: 58% C: 0% D: 28%
8   A: 22% B: 20% C: 21% D: 37%
9   A: 0% B: 48% C: 30% D: 22%
10 A: 52% B: 9% C: 0% D: 39%
11 A: 9% B: 52% C: 39% D: 0%
12 A: 69% B: 0% C: 0% D: 31%
13 A: 23% B: 0% C: 36% D: 41%
14 A: 14% B: 0% C: 43% D: 43%
15 A: 22% B: 57% C: 8% D: 13%
16 A: 4% B: 0% C: 72% D: 24%
17 A: 73% B: 25% C: 2% D: 0%
18 A: 0% B: 19% C: 12% D: 69%
19 A: 16% B: 13% C: 65% D: 6%
20 A: 0% B: 72% C: 5% D: 23%
21 A: 0% B: 28% C: 69% D: 3%

22 A: 9% B: 7% C: 50% D: 34%
23 A: 21% B: 0% C: 0% D: 79%
24 A: 0% B: 43% C: 57% D: 0%
25 A: 0% B: 15% C: 69% D: 15%
26 A: 21% B: 36% C: 29% D: 14%
27 A: 0% B: 92% C: 8% D: 0%
28 A: 12% B: 88% C: 0% D: 0%
29 A: 0% B: 62% C: 0% D: 38%
30 A: 43% B: 36% C: 0% D: 21%
31 A: 4% B: 0% C: 25% D: 71%
32 A: 5% B: 0% C: 16% D: 79%
33 A: 0% B: 92% C: 0% D: 8%
34 A: 9% B: 32% C: 9% D: 50%
35 A: 57% B: 33% C: 10% D: 0%
36 A: 0% B: 19% C: 10% D: 71%
37 A: 66% B: 21% C: 13% D: 0%
38 A: 0% B: 75% C: 6% D: 19%
39 A: 7% B: 12% C: 50% D: 31%
40 A: 0% B: 93% C: 0% D: 7%
41 A: 0% B: 21% C: 15% D: 64%
42 A: 2% B: 19% C: 59% D: 20%
43 A: 30% B: 0% C: 35% D: 35%
44 A: 72% B: 11% C: 0% D: 17%
45 A: 10% B: 61% C: 15% D: 14%
46 A: 5% B: 56% C: 10% D: 29%
47 A: 23% B: 51% C: 20% D: 6%
48 A: 0% B: 62% C: 30% D: 8%
49 A: 24% B: 43% C: 27% D: 6%
50 A: 78% B: 13% C: 9% D: 0%

ASK THE AUDIENCE

£32,000

51 A: 22% B: 58% C: 8% D: 12%

52 A: 7% B: 36% C: 53% D: 4%

53 A: 60% B: 7% C: 19% D: 14%

54 A: 0% B: 93% C: 7% D: 0%

55 A: 75% B: 6% C: 19% D: 0%

56 A: 0% B: 13% C: 63% D: 24%

57 A: 34% B: 6% C: 0% D: 60%

58 A: 0% B: 0% C: 8% D: 92%

59 A: 0% B: 5% C: 16% D: 79%

60 A: 12% B: 0% C: 0% D: 88%

61 A: 72% B: 17% C: 11% D: 0%

62 A: 0% B: 21% C: 43% D: 36%

63 A: 0% B: 69% C: 28% D: 3%

64 A: 78% B: 13% C: 9% D: 0%

65 A: 35% B: 30% C: 0% D: 35%

66 A: 0% B: 88% C: 12% D: 0%

67 A: 32% B: 50% C: 9% D: 9%

68 A: 37% B: 21% C: 20% D: 22%

69 A: 13% B: 0% C: 66% D: 21%

70 A: 9% B: 50% C: 34% D: 7%

71 A: 0% B: 0% C: 12% D: 88%

72 A: 48% B: 30% C: 22% D: 0%

73 A: 11% B: 0% C: 72% D: 17%

74 A: 5% B: 0% C: 79% D: 16%

75 A: 14% B: 61% C: 10% D: 15%

£64,000

1 A: 2% B: 21% C: 61% D: 16%

2 A: 13% B: 38% C: 38% D: 11%

3 A: 63% B: 4% C: 14% D: 19%

4 A: 11% B: 3% C: 25% D: 61%

5 A: 15% B: 8% C: 67% D: 10%

6 A: 14% B: 59% C: 22% D: 5%

7 A: 19% B: 5% C: 12% D: 64%

8 A: 9% B: 41% C: 27% D: 23%

9 A: 0% B: 72% C: 10% D: 18%

10 A: 4% B: 0% C: 6% D: 90%

11 A: 4% B: 0% C: 15% D: 81%

12 A: 79% B: 3% C: 11% D: 7%

13 A: 14% B: 1% C: 5% D: 80%

14 A: 0% B: 77% C: 8% D: 15%

15 A: 12% B: 0% C: 88% D: 0%

16 A: 12% B: 55% C: 26% D: 7%

17 A: 73% B: 4% C: 23% D: 0%

18 A: 83% B: 9% C: 0% D: 8%

19 A: 11% B: 0% C: 5% D: 84%

20 A: 0% B: 91% C: 0% D: 9%

21 A: 0% B: 81% C: 4% D: 15%

22 A: 72% B: 18% C: 10% D: 0%

23 A: 0% B: 69% C: 7% D: 24%

24 A: 2% B: 41% C: 34% D: 23%

25 A: 0% B: 73% C: 27% D: 0%

26 A: 86% B: 9% C: 5% D: 0%

27 A: 46% B: 28% C: 26% D: 0%

28 A: 5% B: 14% C: 45% D: 36%

29 A: 6% B: 15% C: 3% D: 76%

30 A: 27% B: 64% C: 5% D: 4%

31 A: 9% B: 59% C: 19% D: 13%

ASK THE AUDIENCE

£64,000

32 A: 0% B: 23% C: 4% D: 73%
33 A: 0% B: 84% C: 11% D: 5%
34 A: 44% B: 15% C: 41% D: 0%
35 A: 27% B: 0% C: 5% D: 68%
36 A: 13% B: 9% C: 14% D: 64%
37 A: 0% B: 76% C: 19% D: 5%
38 A: 15% B: 13% C: 54% D: 18%
39 A: 7% B: 71% C: 2% D: 20%
40 A: 13% B: 31% C: 29% D: 27%
41 A: 9% B: 0% C: 82% D: 9%
42 A: 36% B: 0% C: 5% D: 59%
43 A: 48% B: 32% C: 10% D: 10%
44 A: 16% B: 27% C: 42% D: 15%
45 A: 4% B: 55% C: 22% D: 19%
46 A: 0% B: 50% C: 23% D: 27%
47 A: 71% B: 4% C: 20% D: 5%
48 A: 0% B: 26% C: 28% D: 46%
49 A: 5% B: 1% C: 14% D: 80%
50 A: 90% B: 6% C: 4% D: 0%

£125,000

1 A: 5% B: 42% C: 34% D: 19%
2 A: 50% B: 7% C: 24% D: 19%
3 A: 0% B: 29% C: 22% D: 49%
4 A: 4% B: 5% C: 79% D: 12%
5 A: 50% B: 21% C: 25% D: 4%
6 A: 19% B: 27% C: 11% D: 43%
7 A: 38% B: 36% C: 22% D: 4%
8 A: 42% B: 18% C: 31% D: 9%

9 A: 20% B: 27% C: 42% D: 11%
10 A: 8% B: 54% C: 6% D: 32%
11 A: 38% B: 4% C: 33% D: 25%
12 A: 13% B: 4% C: 9% D: 74%
13 A: 17% B: 8% C: 54% D: 21%
14 A: 4% B: 4% C: 70% D: 22%
15 A: 4% B: 0% C: 75% D: 21%
16 A: 81% B: 0% C: 0% D: 19%
17 A: 0% B: 65% C: 33% D: 2%
18 A: 29% B: 29% C: 24% D: 18%
19 A: 15% B: 12% C: 42% D: 31%:
20 A: 25% B: 34% C: 4% D: 37%
21 A: 50% B: 25% C: 17% D: 8%
22 A: 46% B: 0% C: 46% D: 8%
23 A: 79% B: 10% C: 11% D: 0%
24 A: 17% B: 8% C: 62% D: 13%
25 A: 12% B: 20% C: 55% D: 13%
26 A: 23% B: 0% C: 65% D: 12%
27 A: 21% B: 9% C: 49% D: 21%
28 A: 81% B: 13% C: 6% D: 0%
29 A: 13% B: 25% C: 54% D: 8%
30 A: 8% B: 25% C: 13% D: 54%
31 A: 5% B: 79% C: 7% D: 9%
32 A: 15% B: 4% C: 2% D: 79%
33 A: 19% B: 8% C: 50% D: 23%
34 A: 21% B: 0% C: 52% D: 27%
35 A: 46% B: 38% C: 10% D: 6%
36 A: 13% B: 21% C: 54% D: 12%
37 A: 24% B: 13% C: 4% D: 59%

ASK THE AUDIENCE

£125,000

38 A: 26% B: 10% C: 3% D: 61%

39 A: 12% B: 0% C: 63% D: 25%

40 A: 21% B: 19% C: 31% D: 29%

41 A: 8% B: 67% C: 15% D: 10%

42 A: 17% B: 10% C: 62% D: 11%

43 A: 17% B: 10% C: 62% D: 11%

44 A: 5% B: 53% C: 12% D: 30%

45 A: 5% B: 42% C: 19% D: 34%

46 A: 50% B: 24% C: 19% D: 7%

47 A: 10% B: 26% C: 3% D: 61%

48 A: 25% B: 4% C: 33% D: 38%

49 A: 27% B: 21% C: 52% D: 0%

50 A: 0% B: 12% C: 25% D: 63%

£250,000

1 A: 20% B: 5% C: 34% D: 41%

2 A: 14% B: 47% C: 16% D: 23%

3 A: 8% B: 52% C: 25% D: 15%

4 A: 43% B: 41% C: 16% D: 0%

5 A: 0% B: 73% C: 24% D: 3%

6 A: 33% B: 39% C: 24% D: 4%

7 A: 35% B: 8% C: 10% D: 47%

8 A: 12% B: 6% C: 43% D: 39%

9 A: 83% B: 14% C: 3% D: 0%

10 A: 32% B: 7% C: 39% D: 22%

11 A: 10% B: 53% C: 33% D: 4%

12 A: 14% B: 5% C: 61% D: 20%

13 A: 0% B: 82% C: 18% D: 0%

14 A: 0% B: 0% C: 23% D: 77%

15 A: 67% B: 6% C: 8% D: 19%

16 A: 15% B: 37% C: 29% D: 19%

17 A: 0% B: 43% C: 30% D: 27%

18 A: 3% B: 47% C: 41% D: 9%

19 A: 2% B: 39% C: 31% D: 28%

20 A: 35% B: 28% C: 23% D: 14%

21 A: 15% B: 11% C: 55% D: 19%

22 A: 6% B: 22% C: 5% D: 67%

23 A: 7% B: 51% C: 9% D: 33%

24 A: 0% B: 50% C: 22% D: 28%

25 A: 18% B: 25% C: 45% D: 12%

26 A: 44% B: 12% C: 0% D: 44%

27 A: 72% B: 9% C: 11% D: 8%

28 A: 17% B: 66% C: 4% D: 13%

29 A: 18% B: 61% C: 5% D: 16%

30 A: 7% B: 41% C: 23% D: 29%

31 A: 0% B: 85% C: 13% D: 2%

32 A: 3% B: 25% C: 39% D: 33%

33 A: 6% B: 28% C: 55% D: 11%

34 A: 27% B: 39% C: 15% D: 19%

35 A: 13% B: 50% C: 4% D: 33%

36 A: 47% B: 10% C: 35% D: 8%

37 A: 19% B: 37% C: 28% D: 16%

38 A: 18% B: 5% C: 61% D: 16%

39 A: 5% B: 18% C: 73% D: 4%

40 A: 95% B: 3% C: 2% D: 0%

41 A: 29% B: 4% C: 5% D: 62%

£250,000

42 A: 80% B: 10% C: 0% D: 10%

43 A: 95% B: 5% C: 0% D: 0%

44 A: 20% B: 0% C: 68% D: 12%

45 A: 59% B: 10% C: 13% D: 18%

46 A: 73% B: 24% C: 3% D: 0%

47 A: 5% B: 10% C: 81% D: 4%

48 A: 24% B: 41% C: 27% D: 8%

49 A: 18% B: 73% C: 4% D: 5%

50 A: 59% B: 10% C: 13% D: 18%

£500,000

1 A: 5% B: 20% C: 41% D: 34%

2 A: 16% B: 47% C: 14% D: 23%

3 A: 52% B: 25% C: 15% D: 8%

4 A: 0% B: 43% C: 41% D: 16%

5 A: 3% B: 0% C: 73% D: 24%

6 A: 5% B: 0% C: 4% D: 91%

7 A: 10% B: 18% C: 59% D: 13%

8 A: 23% B: 19% C: 13% D: 45%

9 A: 59% B: 27% C: 9% D: 5%

10 A: 73% B: 4% C: 5% D: 18%

11 A: 12% B: 0% C: 68% D: 20%

12 A: 15% B: 32% C: 9% D: 44%

13 A: 82% B: 13% C: 1% D: 4%

14 A: 44% B: 10% C: 19% D: 27%

15 A: 24% B: 9% C: 40% D: 27%

16 A: 4% B: 5% C: 29% D: 62%

17 A: 41% B: 13% C: 15% D: 31%

18 A: 36% B: 33% C: 26% D: 5%

19 A: 16% B: 18% C: 21% D: 45%

20 A: 18% B: 21% C: 45% D: 16%

21 A: 14% B: 18% C: 51% D: 17%

22 A: 51% B: 17% C: 14% D: 18%

23 A: 51% B: 17% C: 14% D: 18%

24 A: 20% B: 18% C: 23% D: 39%

25 A: 3% B: 4% C: 85% D: 8%

26 A: 5% B: 50% C: 35% D: 10%

27 A: 36% B: 0% C: 59% D: 5%

28 A: 41% B: 27% C: 8% D: 24%

29 A: 0% B: 72% C: 14% D: 14%

30 A: 0% B: 43% C: 32% D: 25%

31 A: 23% B: 18% C: 23% D: 36%

32 A: 0% B: 0% C: 91% D: 9%

33 A: 10% B: 80% C: 0% D: 10%

34 A: 7% B: 2% C: 86% D: 5%

35 A: 8% B: 14% C: 1% D: 77%

36 A: 14% B: 1% C: 77% D: 8%

37 A: 23% B: 15% C: 3% D: 59%

38 A: 5% B: 0% C: 95% D: 0%

39 A: 10% B: 81% C: 4% D: 5%

40 A: 41% B: 5% C: 9% D: 45%

41 A: 32% B: 66% C: 0% D: 2%

42 A: 16% B: 64% C: 16% D: 4%

43 A: 18% B: 0% C: 65% D: 17%

44 A: 5% B: 0% C: 68% D: 27%

45 A: 10% B: 10% C: 0% D: 80%

ASK THE AUDIENCE

£500,000

46 A: 43% B: 41% C: 16% D: 0%
47 A: 68% B: 20% C: 12% D: 0%
48 A: 0% B: 66% C: 32% D: 2%
49 A: 16% B: 18% C: 21% D: 45%
50 A: 5% B: 0% C: 91% D: 4%

£1,000,000

1 A: 0% B: 60% C: 29% D: 11%
2 A: 2% B: 35% C: 0% D: 63%
3 A: 44% B: 25% C: 13% D: 18%
4 A: 26% B: 3% C: 58% D: 13%
5 A: 59% B: 2% C: 37% D: 2%
6 A: 27% B: 12% C: 5% D: 56%
7 A: 0% B: 63% C: 28% D: 9%
8 A: 34% B: 7% C: 10% D: 49%
9 A: 14% B: 45% C: 29% D: 12%
10 A: 25% B: 8% C: 0% D: 67%
11 A: 43% B: 38% C: 19% D: 0%
12 A: 17% B: 40% C: 29% D: 14%
13 A: 24% B: 0% C: 46% D: 30%
14 A: 15% B: 18% C: 25% D: 42%
15 A: 35% B: 21% C: 8% D: 36%
16 A: 18% B: 37% C: 20% D: 25%
17 A: 33% B: 29% C: 0% D: 38%
18 A: 29% B: 16% C: 32% D: 23%
19 A: 6% B: 67% C: 6% D: 21%
20 A: 46% B: 29% C: 14% D: 11%
21 A: 12% B: 4% C: 19% D: 65%
22 A: 42% B: 16% C: 0% D: 42%

23 A: 18% B: 3% C: 25% D: 54%
24 A: 4% B: 59% C: 32% D: 5%
25 A: 33% B: 6% C: 23% D: 38%
26 A: 7% B: 13% C: 18% D: 62%
27 A: 50% B: 19% C: 12% D: 19%
28 A: 21% B: 21% C: 12% D: 46%
29 A: 0% B: 57% C: 9% D: 34%
30 A: 0% B: 47% C: 22% D: 31%
31 A: 54% B: 25% C: 13% D: 0%
32 A: 13% B: 25% C: 54% D: 8%
33 A: 71% B: 24% C: 5% D: 0%
34 A: 4% B: 5% C: 32% D: 59%
35 A: 18% B: 15% C: 42% D: 25%
36 A: 17% B: 18% C: 51% D: 14%
37 A: 5% B: 32% C: 4% D: 59%
38 A: 31% B: 13% C: 41% D: 15%
39 A: 12% B: 19% C: 65% D: 4%
40 A: 9% B: 57% C: 34% D: 0%
41 A: 6% B: 67% C: 6% D: 21%
42 A: 18% B: 42% C: 15% D: 25%
43 A: 13% B: 25% C: 0% D: 54%
44 A: 21% B: 67% C: 6% D: 6%
45 A: 21% B: 9% C: 55% D: 15%
46 A: 47% B: 31% C: 0% D: 22%
47 A: 63% B: 17% C: 10% D: 10%
48 A: 25% B: 18% C: 44% D: 13%
49 A: 0% B: 38% C: 33% D: 29%
50 A: 2% B: 0% C: 63% D: 35%

ANSWERS

FASTEST FINGER

1	CBDA	21	BCAD	41	BDAC	61	DACB	81	BADC
2	BCAD	22	ABDC	42	CBAD	62	BDAC	82	CADB
3	DACB	23	BCAD	43	BDAC	63	DABC	83	CBDA
4	CBAD	24	CBAD	44	BACD	64	DACB	84	DCAB
5	DBAC	25	BCDA	45	DCBA	65	DCAB	85	BCDA
6	CDAB	26	DBCA	46	CABD	66	DBAC	86	CDAB
7	CDAB	27	CDBA	47	BACD	67	CDAB	87	DBCA
8	DCBA	28	BDAC	48	BADC	68	BCAD	88	CBAD
9	BDAC	29	CADB	49	ACBD	69	DCBA	89	BCDA
10	CBAD	30	DACB	50	DBAC	70	CBAD	90	BACD
11	BADC	31	CABD	51	BACD	71	DACB	91	CBAD
12	DACB	32	BCAD	52	BDAC	72	BDCA	92	DCBA
13	BDAC	33	BACD	53	ADCB	73	BADC	93	BADC
14	ACDB	34	CBAD	54	DBAC	74	DACB	94	BCAD
15	DABC	35	BADC	55	DCAB	75	CDAB	95	BDAC
16	ABDC	36	BADC	56	CDBA	76	CDBA	96	BDCA
17	BADC	37	DBAC	57	ABDC	77	DABC	97	DCBA
18	BCDA	38	CADB	58	CABD	78	ADBC	98	DABC
19	BCAD	39	BCDA	59	ADBC	79	ACDB	99	ADCB
20	CDBA	40	DBAC	60	CABD	80	BDCA	100	ADCB

£100

1	D	8	A	15	C	22	A	29	B
2	C	9	D	16	A	23	C	30	A
3	B	10	B	17	B	24	B	31	C
4	D	11	C	18	B	25	A	32	B
5	C	12	A	19	C	26	A	33	D
6	D	13	B	20	A	27	A	34	C
7	C	14	D	21	D	28	B	35	B

ANSWERS

36 A	49 B	62 B	75 D	88 A
37 A	50 A	63 B	76 C	89 B
38 B	51 C	64 C	77 D	90 A
39 A	52 B	65 B	78 D	91 C
40 A	53 B	66 B	79 A	92 C
41 C	54 A	67 B	80 B	93 B
42 C	55 C	68 C	81 C	94 A
43 B	56 D	69 C	82 B	95 A
44 A	57 D	70 C	83 D	96 C
45 C	58 B	71 A	84 D	97 C
46 B	59 A	72 A	85 C	98 A
47 A	60 A	73 C	86 B	99 B
48 B	61 A	74 C	87 C	100 A

£200

1 A	15 B	29 B	43 A	57 A
2 C	16 A	30 D	44 C	58 C
3 C	17 D	31 D	45 C	59 D
4 D	18 D	32 B	46 B	60 A
5 C	19 A	33 A	47 A	61 A
6 A	20 D	34 B	48 A	62 C
7 B	21 A	35 B	49 C	63 B
8 C	22 D	36 A	50 A	64 D
9 C	23 A	37 B	51 A	65 A
10 C	24 C	38 D	52 C	66 B
11 C	25 A	39 D	53 D	67 A
12 D	26 A	40 B	54 D	68 B
13 D	27 C	41 C	55 C	69 C
14 C	28 A	42 B	56 B	70 B

ANSWERS

71 C	77 A	83 A	89 C	95 B
72 B	78 C	84 C	90 A	96 A
73 B	79 C	85 B	91 A	97 A
74 A	80 D	86 B	92 D	98 B
75 B	81 A	87 A	93 C	99 B
76 D	82 A	88 C	94 D	100 B

£300

1 B	21 D	41 D	61 B	81 C
2 B	22 A	42 B	62 B	82 C
3 B	23 D	43 A	63 C	83 C
4 D	24 B	44 B	64 A	84 D
5 B	25 A	45 D	65 B	85 A
6 C	26 B	46 B	66 B	86 A
7 C	27 A	47 B	67 D	87 B
8 D	28 B	48 C	68 C	88 C
9 C	29 C	49 B	69 C	89 D
10 B	30 C	50 C	70 B	90 C
11 B	31 D	51 D	71 C	91 B
12 C	32 C	52 B	72 C	92 A
13 D	33 D	53 D	73 B	93 C
14 B	34 D	54 C	74 D	94 C
15 C	35 D	55 B	75 D	95 C
16 B	36 C	56 A	76 A	96 B
17 D	37 C	57 B	77 B	97 B
18 B	38 C	58 B	78 A	98 B
19 D	39 D	59 A	79 A	99 D
20 A	40 A	60 C	80 D	100 C

ANSWERS

£500

1	C	21	C	41	C	61	D	81	B
2	A	22	D	42	B	62	C	82	A
3	D	23	B	43	A	63	D	83	A
4	D	24	D	44	C	64	C	84	B
5	B	25	C	45	D	65	A	85	D
6	D	26	B	46	C	66	C	86	C
7	C	27	B	47	B	67	D	87	B
8	A	28	C	48	A	68	B	88	A
9	C	29	D	49	A	69	D	89	B
10	B	30	D	50	C	70	C	90	D
11	B	31	A	51	C	71	C	91	C
12	A	32	A	52	B	72	B	92	B
13	D	33	B	53	D	73	A	93	B
14	D	34	B	54	B	74	D	94	B
15	C	35	C	55	B	75	D	95	C
16	B	36	C	56	C	76	A	96	B
17	C	37	D	57	D	77	D	97	D
18	D	38	B	58	C	78	B	98	D
19	C	39	C	59	A	79	D	99	D
20	C	40	C	60	D	80	B	100	D

£1,000

1	B	7	C	13	D	19	C	25	D
2	D	8	D	14	D	20	C	26	D
3	D	9	A	15	B	21	B	27	C
4	B	10	B	16	A	22	C	28	C
5	C	11	D	17	C	23	C	29	A
6	B	12	A	18	A	24	C	30	B

ANSWERS

31	B	45	B	59	D	73	B	87	B
32	B	46	C	60	B	74	D	88	C
33	B	47	B	61	C	75	C	89	A
34	B	48	A	62	D	76	A	90	C
35	D	49	D	63	C	77	B	91	C
36	B	50	B	64	B	78	C	92	B
37	C	51	D	65	D	79	B	93	A
38	B	52	C	66	A	80	D	94	C
39	C	53	B	67	C	81	D	95	C
40	B	54	B	68	C	82	A	96	C
41	C	55	C	69	A	83	B	97	B
42	A	56	D	70	C	84	D	98	A
43	A	57	B	71	C	85	B	99	C
44	A	58	A	72	B	86	C	100	A

£2,000

1	B	13	A	25	C	37	C	49	A
2	B	14	C	26	B	38	D	50	A
3	D	15	C	27	A	39	C	51	B
4	A	16	A	28	D	40	D	52	B
5	C	17	D	29	C	41	A	53	C
6	C	18	C	30	D	42	C	54	C
7	A	19	A	31	B	43	A	55	A
8	A	20	A	32	B	44	D	56	A
9	A	21	C	33	D	45	B	57	B
10	A	22	B	34	D	46	A	58	B
11	C	23	A	35	A	47	A	59	B
12	D	24	B	36	A	48	B	60	C

ANSWERS

61	D	64	B	67	D	70	C	73	C
62	D	65	C	68	A	71	D	74	D
63	C	66	B	69	C	72	A	75	A

£4,000

1	B	16	C	31	D	46	D	61	A
2	B	17	C	32	A	47	B	62	D
3	A	18	C	33	A	48	A	63	D
4	C	19	A	34	D	49	B	64	D
5	C	20	B	35	B	50	D	65	A
6	C	21	C	36	C	51	C	66	C
7	C	22	B	37	B	52	C	67	D
8	B	23	C	38	D	53	B	68	B
9	D	24	D	39	D	54	C	69	D
10	A	25	C	40	C	55	B	70	A
11	D	26	B	41	B	56	C	71	A
12	C	27	C	42	C	57	B	72	A
13	B	28	D	43	C	58	B	73	D
14	A	29	D	44	B	59	B	74	D
15	B	30	A	45	D	60	C	75	C

£8,000

1	C	8	B	15	D	22	A	29	A
2	A	9	A	16	A	23	B	30	C
3	C	10	B	17	D	24	C	31	B
4	B	11	B	18	B	25	D	32	A
5	B	12	D	19	A	26	B	33	D
6	C	13	A	20	B	27	A	34	D
7	C	14	A	21	D	28	C	35	A

ANSWERS

36	C	44	A	52	C	60	B	68	A
37	C	45	A	53	D	61	D	69	D
38	D	46	A	54	B	62	C	70	D
39	B	47	B	55	A	63	A	71	B
40	D	48	A	56	B	64	A	72	C
41	C	49	B	57	C	65	C	73	A
42	D	50	C	58	D	66	A	74	A
43	C	51	D	59	D	67	B	75	C

£16,000

1	C	16	D	31	B	46	B	61	A
2	D	17	A	32	D	47	B	62	C
3	C	18	A	33	A	48	A	63	C
4	B	19	A	34	A	49	C	64	B
5	C	20	C	35	D	50	C	65	D
6	B	21	B	36	C	51	D	66	D
7	B	22	A	37	D	52	B	67	B
8	B	23	B	38	D	53	D	68	B
9	A	24	B	39	D	54	A	69	A
10	C	25	D	40	C	55	B	70	A
11	B	26	D	41	D	56	A	71	D
12	A	27	A	42	C	57	C	72	B
13	B	28	D	43	D	58	B	73	A
14	C	29	C	44	C	59	D	74	B
15	B	30	D	45	D	60	B	75	D

£32,000

1	D	3	A	5	D	7	B	9	B
2	D	4	B	6	C	8	D	10	A

ANSWERS

11	B	24	C	37	A	50	A	63	B
12	A	25	C	38	B	51	B	64	A
13	D	26	B	39	C	52	C	65	A
14	C	27	B	40	B	53	A	66	B
15	B	28	B	41	D	54	B	67	B
16	C	29	B	42	C	55	A	68	A
17	A	30	A	43	C	56	C	69	C
18	D	31	D	44	A	57	D	70	B
19	C	32	D	45	B	58	D	71	D
20	B	33	B	46	B	59	D	72	A
21	C	34	D	47	B	60	D	73	C
22	C	35	A	48	B	61	A	74	C
23	D	36	D	49	B	62	C	75	B

£64,000

1	C	11	D	21	B	31	B	41	C
2	B	12	A	22	A	32	D	42	D
3	A	13	D	23	B	33	B	43	A
4	D	14	B	24	B	34	A	44	C
5	C	15	C	25	B	35	D	45	B
6	B	16	B	26	A	36	D	46	B
7	D	17	A	27	A	37	B	47	A
8	B	18	A	28	C	38	C	48	D
9	B	19	D	29	D	39	B	49	D
10	D	20	B	30	B	40	B	50	A

£125,000

1	B	3	D	5	A	7	A	9	C
2	A	4	C	6	D	8	A	10	B

ANSWERS

11	A	19	C	27	C	35	A	43	C
12	D	20	D	28	A	36	C	44	B
13	C	21	A	29	C	37	D	45	B
14	C	22	A	30	D	38	D	46	A
15	C	23	A	31	B	39	C	47	D
16	A	24	C	32	D	40	C	48	D
17	B	25	C	33	C	41	B	49	C
18	A	26	C	34	C	42	C	50	D

£250,000

1	D	11	B	21	C	31	B	41	D
2	B	12	C	22	D	32	C	42	A
3	B	13	B	23	B	33	C	43	A
4	A	14	D	24	B	34	B	44	C
5	B	15	A	25	C	35	B	45	A
6	B	16	B	26	D	36	A	46	A
7	D	17	B	27	A	37	B	47	C
8	C	18	B	28	B	38	C	48	B
9	A	19	B	29	B	39	C	49	B
10	C	20	A	30	B	40	A	50	A

£500,000

1	C	8	D	15	C	22	A	29	B
2	B	9	A	16	D	23	A	30	B
3	A	10	A	17	A	24	D	31	D
4	B	11	C	18	A	25	C	32	C
5	C	12	D	19	D	26	B	33	B
6	D	13	A	20	C	27	C	34	C
7	C	14	A	21	C	28	A	35	D

ANSWERS

36	C	39	B	42	B	45	D	48	B
37	D	40	D	43	C	46	A	49	D
38	C	41	B	44	C	47	A	50	C

£1,000,000

1	B	11	A	21	D	31	A	41	B
2	D	12	B	22	D	32	C	42	B
3	A	13	C	23	D	33	A	43	D
4	C	14	D	24	B	34	D	44	B
5	A	15	D	25	D	35	C	45	C
6	D	16	B	26	D	36	C	46	A
7	B	17	D	27	A	37	D	47	A
8	D	18	C	28	D	38	C	48	C
9	B	19	B	29	B	39	C	49	B
10	D	20	A	30	B	40	B	50	C